ARCHITECTURALLY EXPOSED STRUCTURAL STEEL

SPECIFICATIONS / CONNECTIONS / DETAILS

TERRI MEYER BOAKE

ARCHITECTURALLY EXPOSED STRUCTURAL STEEL

SPECIFICATIONS / CONNECTIONS / DETAILS

BIRKHÄUSER / BASEL

The preparation of this book was kindly supported by the
World Steel Association.

worldsteel
ASSOCIATION

GRAPHIC DESIGN, LAYOUT AND TYPOGRAPHY
Reinhard Steger, Christian Schärmer, Deborah van Mourik
Proxi, Barcelona

EDITOR FOR THE PUBLISHER
Andreas Müller, Berlin

LIBRARY OF CONGRESS
CATALOGING-IN-PUBLICATION DATA
A CIP catalog record for this book has been
applied for at the Library of Congress.

BIBLIOGRAPHIC INFORMATION PUBLISHED BY THE
DEUTSCHE NATIONALBIBLIOTHEK
The Deutsche Nationalbibliothek lists this publication in the Deutsche
Nationalbibliografie; detailed bibliographic data are available in the
Internet at http://dnb.dnb.de.

This publication is also available as an
e-book pdf (ISBN 978-3-03821-483-0) and
EPUB (978-3-03821-446-5).

P.O. Box 44, 4009 Basel, Switzerland
Part of Walter De Gruyter GmbH, Berlin/Munich/Boston

Printed on acid-free paper produced from
chlorine-free pulp. TCF ∞

Printed in Germany

ISBN 978-3-03821-574-5

9 8 7 6 5 4 3 2 1
www.birkhauser.com

CONTENTS

BUILDING PROFILES

municon

FOREWORD

The World Steel Association (worldsteel) is proud to be the exclusive sponsor of *Architecturally Exposed Structural Steel: Specifications, Connections, Details*. Construction is one of the most important steel-using industries, globally accounting for more than 50% of world steel use. Buildings – from houses to car parks to schools to skyscrapers – rely on steel for their strength and durability. In addition to structural frameworks, steel is also used on many other parts of buildings, including roofs and cladding for exterior walls.

Steel continues to be at the root of advances in architecture and construction. The use of exposed steel in buildings brings the design benefits and dynamic potential of steel to the public eye. Its stiffness allows steel to span greater distances and provides more design freedom than other materials. Steel's superior strength-to-weight ratio makes it possible for the structure to bear high loads using less material. Architecturally Exposed Structural Steel (AESS) plays a significant role also in the design of contemporary pedestrian bridges that elevates their role in the urban realm to that of art.

Sustainable steel is at the core of a green economy. Reusing or recycling building components is key to the sustainability of a structure's end-of-life as it is the most economic and ecological solution. The global recovery rates for steel construction applications stand at 85%, making it a good choice for building structures. The exposure of steel leads to a reduction of materials that would otherwise be used to conceal the structural systems, while at the same time creating stimulating architecture.

Steel is safe, innovative and progressive. Industry surveys consistently demonstrate that steel is the safest construction material. Steel offers the highest strength-to-weight ratio of any building material. Because of its strength and durability, steel structures are designed to withstand natural disasters. It is also impervious to attacks from termites or fungi, does not rot or split and is highly fire-resistant. The steel industry globally spends more than €12 billion annually on improving the manufacturing processes, new product developments and future breakthrough technologies.

Steel is a key driver of the world's economy. The industry directly employs more than two million people worldwide, with a further two million contractors and four million people in supporting industries. In 2013, the steel industry had a turnover of more than $900 billion, yielding over $100 billion in tax.

Steel plays a fundamental role in the development of modern societies and is an ideal material to help meet the societies' growing needs for buildings and infrastructures in a sustainable way. Its intrinsic properties such as its strength, versatility, durability and 100% recyclability allow improved environmental performance across the entire life cycle of buildings.

The AESS Category System of design presented in this book acknowledges the importance of the role of proper connection design and erection strategies, and communication between the fabricator, engineer and architect, as central to ensuring safety on the site.

Dr. Edwin Basson
Director General, World Steel Association

The Munich Airport Center in Munich, Germany, designed by Helmut Jahn and completed in 1999, showcases high-level detailing in Architecturally Exposed Structural Steel.

ACKNOWLEDGMENTS

This publication has been made possible through the generous support of worldsteel. Worldsteel was also the sponsor of *Diagrid Structures: Systems, Connections, Details*.

My early interest in steel must be credited to inspiration from the High Tech works of Foster, Rogers, Piano and Grimshaw as well as Santiago Calatrava. My appreciation of the importance of the details of these connections as elements of industrial design was inspired by my former colleague at the University of Waterloo School of Architecture, Ron Keenberg of IKOY Architects.

Much of my energy and understanding of Architecturally Exposed Structural Steel comes through my involvement with the Canadian Institute of Steel Construction in the development of the methodology that is expanded upon in this book. I owe so much to the large number of steel professionals involved in this development. Particular thanks go to Mike Gilmor, Rob Third and Suja John.

Also core to the CISC AESS committee work, my understanding and experience of steel would be nowhere without the assistance of Sylvie Boulanger, Walter Koppelaar and Tim Verhey. Sylvie is my engineering counterpart and has willingly shared so much knowledge and insight with me. Sylvie traveled with me to Australia and New Zealand in conjunction with expanding the Category System to Australasia. Walter has always encouraged me and allowed me into his fabrication shop, and provided access to numerous job sites and high-quality projects. Without these detailed first-hand experiences of construction in process, my expertise would not have progressed beyond that of a standard instructor and my image bank would be substantially poorer. Tim Verhey was always willing to provide me with very detailed technical clarifications, many of which are included in this book.

Thank you to Alistair Fussell of Steel Construction New Zealand and David Ryan of the Australian Steel Institute for hosting Sylvie and me and providing access to some excellent steel projects, many of which are included in this book.

Thanks as well to my editor Andreas Müller for his expertise, input and support and to Reinhard Steger and his team for their loving attention to the graphic layout of this book.

Lastly to my family for the continued support of and pride in my publication endeavors.

PREFACE

I have found great joy and inspiration in the detailed study of Architecturally Exposed Structural Steel (AESS) that has ensued from the research and writing I undertook for my first book *Understanding Steel Design: An Architectural Design Manual*, published by Birkhäuser in 2011. Where *Understanding Steel Design* looked more holistically at the potential in the informed use of steel in architectural projects, *Architecturally Exposed Structural Steel* looks more specifically at an approach to properly designing steel for exposure. Such exposure demands that architects become more engaged in the detailed design of steel systems, and with that, understand better the varying approaches to connection design, fabrication and finishing and how these impact the entire project in significant ways.

The category approach to specifying and detailing AESS that is presented in this book is derived from substantial work done in conjunction with the Canadian Institute of Steel Construction (CISC) between 2005 and 2011. The development of this system by CISC was intended to create better communications within projects in order to make the incorporation of AESS more straightforward by standardizing some of the core fabrication and detailing practices, thereby allowing the team to focus on the "real questions". The method has subsequently been adopted in New Zealand and Australia and I am working with a large team in association with the American Institute of Steel Construction (AISC) to create a unified approach to AESS practices additionally between the USA and Canada.

The photos included in this book were predominantly taken by myself during building visits. As the design of AESS strongly focuses on the details of finished steel, it was essential that the derivation of the Category System be done through first-hand experience, as it is the best means to validate the characteristics of projects. Certain project images taken during fabrication processes where I was unable to be present, have been sourced from the team members. These are credited at the back of this book.

Ronald Reagan National Airport in Washington, DC, USA, designed by Pelli Clarke Pelli Architects, demonstrates the possibilities of AESS using a highly angular aesthetic. Here the curves are faceted. Much of the high-level steel makes use of simple structural shapes. A combination of remediated welded connections is mixed with exposed bolting. The extensive use of natural light successfully animates the technical appearance of the steel.

THE BASIS OF ARCHITECTURALLY EXPOSED STRUCTURAL STEEL

CHAPTER 1

WHAT IS AESS?

Architecturally Exposed Structural Steel (AESS) is steel that must meet two requirements: it must be designed to be structurally sufficient to support the primary needs of the structure of the building, canopies or ancillary structures, while at the same time being exposed to view and forming a significant part of the architectural language of the building. Any structural steel that is not concealed can therefore be considered architecturally exposed. The design, detailing and finish requirements of AESS will typically exceed that of standard structural steel, which is normally concealed by other materials or finishes. This naturally increases the time and cost to design, detail, fabricate, erect and finish AESS systems.

Perth Arena in Perth, Australia, designed by ARM Architecture and CCN, uses finely welded round HSS tubes to create a dynamic entrance structure for this arena completed in 2012. The complex angles that challenge the fabrication of the structure push this construction to an AESS 4 category.

The categorization of structural steel as architecturally exposed necessitates a new approach to its design and detailing, in particular as not all AESS need be designed to be equal(ly expensive). Retail stores, arenas, museums and airports will each have very different expectations of the role of the steel in the aesthetics of the building. Even within building typologies, different approaches to steel design are valid.

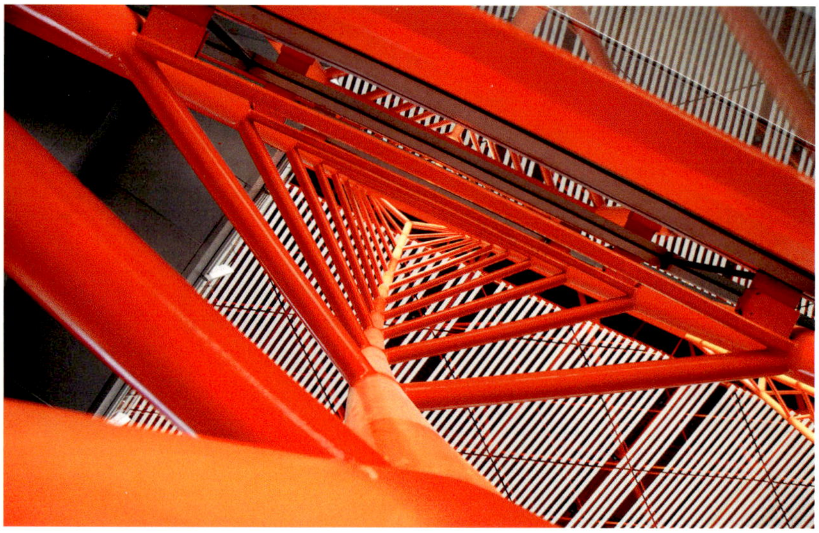

Heathrow Airport Terminal 5 in London, England, designed by Rogers Stirk Harbour + Partners, Arup and Watson Steel Structures limited (top left), is an example of very high-end custom fabrication and detailing. The connections make use of bolting and welding as appropriate to fabrication and erection and allow the process of the assembly to be seen.

The detailing of the Canadian Museum for Human Rights in Winnipeg, MN, Canada, designed by Antoine Predock (bottom left), makes use of a combination of custom-fabricated elements and more standard sections and detailing. Bolting is proactively used for its rugged aesthetic as part of the story told by the materiality of this museum.

Beijing Capital International Airport Terminal 3 in Beijing, China, designed by Foster + Partners, with Arup (top) is an example of high-end AESS that looks for a very smooth, all-welded solution. Connections that are more technical in character and might use bolting are hidden above the white slatted suspended ceiling.

The nature and level of fabrication detailing needs to address also the distance to view. It is unnecessary to detail connections and finishes located beyond normal view (approximately 6m/20ft) in the same way as steel located within range of view and sometimes touch.

The elaborate HSS truss system to support the pyramidal skylight over this city hall atrium in Edmonton, AB, Canada, designed by Dub Architects, is located well above view. It is therefore acceptable that the connections have not been fully remediated. This provided significant cost savings for the project.

Whereas designers tend not to be involved in connection issues for concealed structural systems, exposed systems tend to become the architectural trademark of the building, hence requiring the architects' involvement. Up to this point it has been difficult for designers to be adequately informed for choosing appropriate methods of AESS design and detailing or on the cost implications of their choices.

The surge in the use of AESS has created a paradigm shift in the sequential communication that usually takes place in projects where the steel structure is hidden. The paradigm shift centers on the simple fact that a "well-designed connection" or a "smooth surface" has very different meanings for an architect, an engineer or a fabricator. Such a situation can create a misalignment of communication, especially in terms of what can be accomplished within specific budget limitations.

In most countries, architectural and engineering contract documents do not contain much specification on the requirements for AESS. Exposed steel has normally been considered covered by the structural steel specification, with some additional notes and comments of clarification. As AESS applications become more complex and extensive, this rough method of communication has proven to be very ineffective.

It is the intention of this book to set out a very clear and systematic approach to designing, specifying and detailing AESS. The use of AESS on a project can be simplified by applying a standard, tiered, additive category approach. The requirements for the steel according to the categories are based primarily on the use of the building, the distance of view of the structural elements, aesthetic expectations and budget.

The AESS Specification Category System that forms the basis of the discussion in this book references a methodology that was developed by the Canadian Institute of Steel Construction, which has subsequently been adopted by Australia and New Zealand and is also being used in some projects in the USA. The origins of the tiered category approach go back to an initial document published by the American Institute of Steel Construction in *Modern Steel Construction* in May 2003. The AESS Specification Category System is the most detailed and comprehensive approach to AESS worldwide. This method was previously introduced in the chapter "Architecturally Exposed Structural Steel (AESS): Design and Detailing Requirements" in *Understanding Steel Design: An Architectural Design Manual.*

Although the categories of AESS 1 to AESS 4 have been developed specifically for the Canadian and Australasian systems, the idea of differentiated categories of AESS is generally applicable as they reflect differences in finish and detailing that respect differences in viewing distance, function and cost. The projects in this book are examined according to their potential position in the Category System, regardless of whether the system was actually used in their detailing and fabrication, with the aim to build a visual language that corresponds to the types of details and the fabrication methods that can be used in each of the AESS categories. For more information and specifications of the specific national systems visit: http://www.cisc-icca.ca/solutions-centre/aess.

THE EVOLUTION OF ARCHITECTURALLY EXPOSED STRUCTURAL STEEL

The architecturally motivated exposure of the structural system has its roots in Gothic architecture. Here the stone structure, columns and buttress support systems were detailed expressively and effectively became the aesthetic of this style. Iron and steel systems as they were developed in the 1700s and 1800s continued in the spirit of this approach. A methodology of framed, elemental construction associated with a specific language of connections arose from the 19th-century Structural Rationalist movement. The aesthetics were clearly driven and unfolded by the method of the manufacture of the iron. Cast iron lent itself to the development of more decorative, repetitive elements. Wrought iron elements tended to be more simple due to their manufacturing process. Both construction types used erection methods that centered on the use of physical connections for joining the structural elements into a stable system. As the brittle nature of the chemistry of cast iron negated the use of welding, joinery generally made predominant use of hot rivets, a method that fell out of favor with the invention of high-strength steel and the use of modern welding and bolting.

Steel detailing and connection design is intrinsically tied back to the types of profiles that are being connected. Early steel manufacturing processes in the 1800s were only able to fabricate angles, plates and later I-beam sections. The invention of processes that could hot-roll larger wide-flange or Universal column types generated the standard connection detailing typical in early Modern structures. Although round pipe had been technically available since the early part of the 20th century, the material was cast by nature and predominantly intended for use in plumbing and services. It could not be welded and was not particularly friendly to work with in structural applications. Modern hollow structural sections (HSS) were not invented until the 1970s. These are fabricated from hot-rolled steel and therefore have the same chemical properties as regular sections.

The Bibliothèque Sainte-Geneviève in Paris, France, designed by Henri Labrouste, 1838, is a fine example of Structural Rationalist exposed cast iron detailing. The high level of ornate decoration was achievable by the casting process, which, however, also led to rigid repetition of identical elements and consequently a low level of variability in the design.

The iconic Farnsworth House in Plano, IL, USA, designed by Ludwig Mies van der Rohe, 1945–1951, makes use of simple hot-rolled shapes (channel, wide-flange and angle). The connections were welded to keep the detailing very clean and somewhat "mysterious".

The Sainsbury Centre for Visual Arts in Norwich, England, designed by Foster + Partners, 1977, demonstrates the very specific use of structural sections and connection detailing that distinguishes High Tech from the International Style. Hot-rolled tubular steel was quite new at the time of construction of this project, necessitating new approaches to fabrication and connection design.

The invention of HSS marked a major stylistic break in exposed steel expression. The range of available sizes increased slowly and eventually had an impact on the design of exposed steel structures. Very large diameters and elliptical sections only became available around 2005, closely followed by technical advancements in CAD and structural calculation programs. These assisted in an explosion of possibilities in the application of the materials to buildings with increasingly complex geometries.

FROM STRUCTURAL RATIONALISM TO HIGH TECH

Significant inventions in the recent history of steel construction have resulted in shifts in design and design theory. The geometric simplicity of International Style architecture aligned well with the limited palette of hot-rolled structural sections that were available through the beginning of the 1970s. Steel structures of this period were more often concealed than exposed, while exposed steel tended to be limited to quite standard welded or bolted connections that suited the available wide-flange (Universal) sections. The British High Tech movement was able to exploit the invention of tubular material, whose geometries challenged the articulation and expression of this new language of exposed steel connections. Tubular steel required the complete rethinking of connection design. Many of the designs of this period also incorporated tension and suspension systems. These too added design requirements to the exposure of connection detailing.

The very basis of High Tech design lies in its roots as an "assembled" and largely prefabricated methodology putting unparalleled emphasis on the design of connections. It is about taking "rationalized industrial technology into building construction" (Peter Buchanan).[1] The High Tech movement in England incorporated this focus into architectural manifestation through a transformed reinvention of earlier Structural Rationalist methodology. This industrialized construction system is also responsible for providing steel with a competitive edge over site-cast systems by preferencing shop fabrication that tends to be faster and more economical than site fabrication. The shop environment is safe from the negative impacts of climate and normally able to use less expensive labor.

High Tech design also included the exposure of mechanical services and systems. The associated exposure of the interior applications of steel adds to the coordination requirements as mechanical systems elements are also exposed to view and thereby become a critical part of the architectural aesthetic. The choice must be made to highlight or minimize their influence on the space. This concern has carried directly into considerations in AESS applications.

The mechanical systems on the interior and exterior of the Centre Pompidou in Paris, France, designed by Renzo Piano and Richard Rogers, 1972–1976, required increased coordination, also because the appearance of these systems fed into the expression of the project.

The louver system on the ceiling of the Sainsbury Centre for Visual Arts allows partial exposure of the mechanical systems above and increased the coordination requirements with the exposed steel systems.

Even though High Tech is similar to Structural Rationalism in its exposure of the structural framework and embellishment of the connection details, it more clearly differentiates the members by their tensile or compressive load capabilities. The high tensile capabilities of modern steel enables this force-varied expression. This was not possible in the age of wrought or cast iron as the tensile capabilities were much smaller and engineering science not adequately developed. Common to most High Tech buildings is an expanded use of lightweight tension systems. Many of these projects are about "doing more with more" and created elaborate structural systems to solve problems that could arguably have been constructed more simply. The High Tech buildings do serve to explore the architectural potential of steel and steel connections as expressive devices. They bear revisiting as precedents for contemporary detailing in AESS structures.

INITIAL DEVELOPMENTS IN HIGH TECH DETAILING

The British High Tech movement did much to establish the basis for the detailing of expressed steel connections. As presented in the chapter "AESS: Its History and Development" in *Understanding Steel Design: An Architectural Design Manual*, the High Tech style was rooted in the regularity of the systems and the ease of repetition of the primary elements. Further to this we see buildings that:

- make fairly exclusive use of purpose-fabricated components
- make predominant use of standard, off-the-shelf components
- buildings that use an even mix of specialty and off-the-shelf components
- buildings that place a high level of attention on connection detailing

Early High Tech buildings created a clear distinction between the fabrication processes that were best handled in the shop – welding – and the erection processes that were most efficient on the site – bolting. This holds true both for projects that were making use of entirely custom components and those that incorporated more standard sections; prefabrication was consistently applied in all projects. This critical recognition of process has been carried over into contemporary methods for AESS. The early High Tech projects developed approaches to connection design that continue to influence the way that AESS detailing is approached.

The connection language that characterized 19th-century Structural Rationalist buildings, and subsequently High Tech architecture, continues to be developed, perfected and exploited as one of the desirable aesthetic features of Architecturally Exposed Structural Steel. This type of structural expression demands that architects become increasingly engaged in understanding the design, detailing and construction of steel structures, not only as a function of the engineering of such forms, but also with regard to the realities and promises to be found in their fabrication.

The moment of transition from High Tech to AESS happened around the time of the design of Chicago O'Hare International Airport United Airlines Terminal by Helmut Jahn from 1985 to 1988. Up to that point, the exploration of High Tech had been contained within a small number of predominantly British or other European practices and few of the projects had been constructed outside of the European mainland and the United Kingdom. O'Hare was a remarkable deviation from the established tradition of reinforced concrete use in airport design, represented by Eero Saarinen's Dulles Airport in Washington, DC, and the TWA Terminal at John F. Kennedy Airport in New York City.

High Tech detailing such as explored in the Centre Pompidou needed to address the use of force-differentiated systems. The use of exposed tension systems had to that point not formed a dominant theme in architecture. The geometric resolution of multiple forces through a point, as explored in the Centre Pompidou, was to become an important design focus in the development of AESS systems.

The interior of the United Airlines Terminal at Chicago O'Hare International Airport in Illinois, USA, designed by Helmut Jahn, makes critical use of the expression of the steel detailing as the primary aesthetic of the architecture.

The use of exposed steel at O'Hare was effective in setting a precedent for this rapidly expanding building type. A high percentage of airports constructed after 1988 have chosen to use AESS, some of which are examined in this book, as they have assisted in a change of attitudes towards innovation and detailing.

From the late 1980s to the present time, as more and more exposed steel has been included in the design of buildings, subtle changes have emerged in the language of detailing. Some of the exuberance and extremely expressive detailing in earlier projects has softened. This language of detailing forms the basis of the studies in this book.

PRIMARY FACTORS OF INFLUENCE THAT DEFINE AESS

Architecturally Exposed Structural Steel has design criteria that differ significantly from standard steel framing. High Tech use of exposed steel and early AESS applications tended towards a very high-end product that was typically constructed using predominantly custom-fabricated pieces with complex detailing throughout. The use of exposed steel in High Tech architecture was exceptional at the time as it contrasted with almost all other aesthetic trends, which chose to conceal structural steel. High Tech projects were typically high-profile examples of building typologies with larger budgets. To carry this ideology into a definition of AESS, which includes all exposed steel, has been problematic. It led to a tendency to over-design and over-specify AESS on projects where this was not warranted. It must be recognized that some projects can be priced out of possibility if the expectations on detailing outstrip the project budget.

The differentiation of the requirements of AESS in a sequence of categories arises from an examination of the role that the exposed steel plays in the overall design, as a function of the building type and its spatial geometries. As AESS is a key element of the expression of the architectural design, emphasis is on the form of the steel in the design. The overall form may vary greatly from regular framing to include curves, unusual angles or three-dimensional elements. These variations add complexity and cost to fabrication and erection. Members and connections need to be designed with more attention to the way that their details support the aesthetic intentions of the design. Bolted or welded connections may be chosen less for their structural capabilities or ease of erection and more for the way that their appearance works with the overall design intention. This said, their structural integrity must remain a key design consideration.

Highly articulated steel structures are by their nature more difficult to fit. There is significantly less play admissible in the connections and accumulated errors can result in overall misalignment. The need to ensure accuracy, ease of fabrication, as well as bottom-line constructability puts greater pressure on the details and requires narrower tolerances throughout the entire project. Tighter tolerances will carry through when the exposed steel framing must coordinate with other trades, in particular regarding areas of significant glazing and curtain wall. The use of stainless steel spider connections for structural glass systems puts additional pressure on allowable tolerances.

Another factor is the selection of the finish, which must take place already at the beginning of the AESS design process. Finishes of exposed steel vary as a function both of the design intention and of weathering, interior or exterior exposure and fire protection. A high-gloss finish will reveal every imperfection and therefore will require more fastidious fabrication. A thicker intumescent coating will conceal many surface imperfections. Galvanizing has issues with consistency of finish and its selection may accompany a less polished selection of details. The bottom line for the contract is that both time and money will be wasted if the level of fabrication care greatly exceeds the nature of the finish.

The high-end AESS detailing of Brookfield Place in Toronto, ON, Canada, designed by Santiago Calatrava, 1996, was long looked at as a quality benchmark for all AESS detailing. Yet even in this project, the nature of the fabrication and joints is different at the base and at the skylight support system. Welding is used at the lower levels where a continuous, smooth finish is desired; bolting is used at the higher levels where visual access is more limited.

Three primary physical conditions feed into the detailing of an AESS structure:

Viewing Distance: 6m/20ft is chosen as a base dimension of the distance up to which an occupant is able to scrutinize the finish from close range and even touch the product. 6m represents a normal height of a high ceiling. For the design of atrium spaces it is important to apply this measure also to the horizontal and downward dimensions, given that the view across or down a space is as critical as the upward view. Where steel is viewed from above, care must be taken to detail the steel so as to avoid the build-up of grime and trash. Viewing distance can also impact the requirements for the surface finish, as some natural blemishes in the steel from the manufacturing, fabrication or mill processes will not be able to be perceived at a distance.

Building Type or Function: The exposed steel over an ice rink and the exposed steel in an airport are likely to have different aesthetic and finishing requirements. In the same vein, the program of a building and the range of spaces should be examined to assess whether different categories of AESS may be specified. Exposed roof trusses, for example, may be a lower category and the columns or base details a higher category. If this is clearly marked on the contract drawings, then the fabricator can adjust the bid according to the appropriate level of fabrication and finish.

Potential Cost Increase: There is a range of increase in the cost to fabricate and erect AESS over the comparable cost of standard structural steel. Additional time is required in fabrication. The erection costs will also increase as a function of the complexity of the steel, the degree to which complex connections can be fabricated in the shop, transportation, access and staging area concerns and increased tolerance requirements to fit the steel. The more complex the AESS and the higher the nature of the finishing requirements, the tighter the tolerances become. This also increases the time to erect the steel and the cost as the labor portion associated with the material grows.

In order to design and detail AESS in an appropriate way a differentiated approach must therefore recognize the following realities.

- Distance and visibility: If you cannot see it up close or touch it, the finish requirements can be softened.
- Bolted versus welded connections: Different connection types result in different aesthetics, requiring differing levels of finish.
- Tolerances at fabrication and erection: Different tolerances are needed as a function of the scope and complexity of the project, but typically tighter tolerances are needed for AESS than for standard structural steel framing.
- Access to detail to perform the required finish: The greater concern for workmanship may mean altering the detail or its location to allow access for different types of tools. Highly articulated connections or details need to be constructible and maintainable.
- Degree of expression: The complexity of structure and connections will impact the detailing and cost.
- Size and shape of structural elements: W (Universal) and HSS shapes have different connection requirements and their use infers a different approach to detailing and finish.
- Interior or exterior setting: This impacts weathering issues, cleaning and maintenance access, the need for fire protection and the potential for impact damage.
- Paint finish, corrosion resistance, fire protection: Depending on the relative thickness of the finish material, more or less care may be required when preparing the surface, edges and welding of the steel.

COMMUNICATION ISSUES BETWEEN ARCHITECT, ENGINEER AND FABRICATOR

Since the advent of the increased use of AESS over the last four decades, the management, work process and design flow of projects have changed. Traditional concealed steel has become routine in terms of its details and construction practices. Detailing has been managed between the structural engineer and steel fabricator, detailer and erector. Expressed steel design, by contrast, must include the architect and may take the designer into a different role in the development of the project details. Increasingly, the development of the AESS system and connection design and detailing requires the input of the fabricator, a player who has not been part of the traditional project team. It is the steel fabricator who best understands the fabrication and erection processes and who can provide valuable insights into the decision-making process.

A collaborative design dialogue is more critical here than for other structural applications that are concealed. Using AESS necessarily increases the intensity of the dialogue as the member choice and fabrication details become critical in conveying the architectural aesthetic. Effective design with AESS emphatically requires a team approach.

One of the major advantages of the AESS Category System is its ability to focus the dialogue of the team on important issues by establishing a standard level of product. This means that not all aspects of fabrication, detailing and workmanship need to be debated. These are established by the choice of the respective category and become standard expectations as specified by the set of characteristics associated with each of the AESS categories.

THE ROLE OF BIM AND DETAILING SOFTWARE

Technical drawings have changed significantly over the last 30 years. The 20th-century tendency towards the construction of orthogonal steel frame buildings using simple, repetitive systems can in part be attributed to limitations in representation, fabrication and calculation. Most steel frame buildings could be calculated as determinate structures, resolving most force systems into planar ones. Even through the early phase of the High Tech movement, calculations were done using a sliderule and the building was drawn manually. Variations from standard details and systems were considered exceptional and their practice was limited. Even if High Tech architecture has a complex appearance, this was achieved with an extremely high degree of repetition of identical elements.

The development of Building Information Modeling (BIM) has been critical to allowing for the design and fabrication of more complex steel structures, without prohibitively greater commitments to time and expense. Geometric complexity requires a very high level of accuracy and makes the assessment of loading on members and connections far more challenging than rectilinear structures that can be reduced to simple, determinate transfers of load. BIM and programs such as Xsteel were created to assist in engineering, detailing and fabricating the steel that supported these unusual forms. The transition of Xsteel to Tekla Structures in 2004 added significantly more functionality and interoperability to the previously available software. The interoperability of the workflow is essential to a successful collaboration between the architect, engineer and steel fabricator.

Detail drawing produced using Tekla Structures, by Walters Inc., for QRC West in Toronto, ON, Canada, designed by &Co Architects. Current steel detailing software is able to incorporate a wide variety of geometries and allows the project team to visualize the nature of the connections. Such software is able to produce the drawing sets necessary for the fabrication and erection sequencing of the steel.

The structural drawings, such as those produced using Tekla Structures, Bentley Systems and others, perform a critical role in the design and construction process. These drawings are created by the steel fabricator and used to detail all of the connections in the project, based on structural information provided by the engineer. It is possible to zoom into the model to create or examine each and every connection. The 3D model becomes a useful tool in the design process as it permits the team to see and share the connection design. This sort of 3D model can be an effective replacement for very expensive physical mock-ups, which in addition can delay the project delivery. The color format allows the fabricator to clearly identify systems and visualize the erection sequence. The shop drawings that are required for each and every member and connection are extracted from this model. The use of this sort of system helps to ensure a high level of accuracy in the fabrication of the elements, which in turn assists in creating tighter tolerances and better erection.

THE CATEGORY SYSTEM

CHAPTER 2

The underlying principle in the design of AESS is the establishment of categories that are differentiated by their architectural requirements and consequently by the technical and cost requirements. The categorization is fine-tuned according to issues of distance to view, type or function of the building and the required budget. A tiered category approach to specifying AESS is essential for the design and decision-making process for a project. Not all projects will have the same requirements for the steel in terms of fabrication and finishing. It is therefore a waste of expense and time to create very high-quality steel for all projects. The selection of an appropriate category for the steel in the project provides the designer with a firm starting point for design detailing.

The AESS diagrid columns that support the roof of the World Financial Center entry pavilion in New York City, NY, USA, designed by Pelli Clarke Pelli Architects, meet the highest standards of AESS 4 design. Fabrication by Walters Inc. and erection by Metropolitan Walters.

Although the categories of AESS 1 to AESS 4 have been developed specifically for the Canadian and Australasian systems, the idea of differentiated categories of AESS is generally applicable as they reflect differences in finish and detailing that respect differences in viewing distance, function and cost. The projects in this book are examined according to their potential position in the Category System, regardless of whether the system was actually used in their detailing and fabrication, with the aim to build a visual language that corresponds to the types of details and the fabrication methods that can be used in each of the AESS categories. For more information and specifications of the specific national systems visit: http://www.cisc-icca.ca/solutions-centre/aess. This chapter builds on information first introduced in the chapter "Architecturally Exposed Structural Steel (AESS): Design and Detailing Requirements" in *Understanding Steel Design: An Architectural Design Manual.*

The categories of AESS are based on the performance requirements for standard structural steel, which they comprise and exceed for the specific AESS matters. The specification requirements for concealed steel structures are addressed within the structural steel specifications, which are normally in the jurisdiction of the engineer. This is important to note: it is normal practice for steel fabricators to refer to the engineering specifications and drawings for a project, not to the architectural contractual documents. Furthermore, within the categories the category approach is additive. The specific requirements for the steel for a given category are added to the requirements for the previous category. The structural requirements for standard steel provide the base for all.

It is to be noted that the viewing distance is a suggestion and not an absolute rule. It is possible to position AESS 2 steel closer to view than 6m/20ft, with the understanding that the product will not appear as finely finished as with steel of a higher category. This might be suitable as a function of detailing, aesthetics, building function and budget.

TO GRIND OR NOT TO GRIND?

A major motive behind the development of the category-based AESS system is to assist the industry and architects in making good choices. In studying problems associated with the absence of a clear specification for AESS, it became apparent that many projects, in an attempt to exhibit "high-quality finish", were over-specifying the grinding of welds. Such remediation of the intrinsic conditions of a connection method is only warranted on the highest AESS quality levels. It is important to understand that the weld has a functional use and its removal will weaken the connection. Competent and neat welding should suffice on all but the highest AESS categories, with deference to AESS 4. Sufficient distance of view or the specifics of certain building functions will obviate the need to grind welds.

This level of remediation to welding came to be recognized as the major point of differentiation between the lower and upper AESS categories. Grinding of welds is only admissible at AESS 3 and above. In the chapters to follow we will look in detail at connections with this understanding of fabrication practices and how they feed into the design.

There are several issues to be addressed regarding grinding.

- Welds are structurally necessary and designed to join the steel members. If welds are to be ground smooth, a different type of welded connection needs to be specified to avoid removing all of the weld material through grinding.
- Grinding is time-consuming and expensive. To spend the time and budget on grinding welds that are out of view is quite pointless.
- Grinding can be dangerous. Grinding operations are best carried out in the shop, where the steel is more easily turned and manipulated to the advantage of access. If grinding must be done to on-site connections, proper platforms need to be erected to provide the ironworkers with safe access.

A "glove-smooth" finish is only suited to AESS 4 projects. The grinding and filling of surfaces prior to the application of a high-quality paint finish should be considered far from routine. It will incur a high-cost premium and is seldom warranted. Skilled welding that reveals the art of metalworking is quite acceptable in many situations.

STANDARD STRUCTURAL STEEL

The initial point of technical reference for AESS structures is standard structural steel, a material established and well understood as a baseline in construction specifications around the world.

The understanding of categories of Architecturally Exposed Structural Steel begins by differentiating structural steel in terms of its degree of exposure. It is assumed that regular structural steel is normally concealed for reasons of finish preference or because of the selected fire protection method. In normal circumstances, where standard structural steel will be either clad or fire-protected or both, there is little or no architectural concern over the design of the details, connections and even the type of members chosen. Although some applications will be more complicated than others, and hence priced accordingly, standard structural steel is not subject to the same considerations as an exposed product.

The structural integrity of standard structural steel is clearly the overriding concern of this material. Architecturally Exposed Structural Steel will follow all of the same structural requirements and be subject to additional requirements as defined by the assigned AESS category (1, 2, 3, 4 or Custom) and the specific set of characteristics associated with a specific AESS category.

The requirements even of highly complex connections are dominated by structural, not by aesthetic concerns. This connection at the Addition to the Royal Ontario Museum in Toronto, ON, Canada, designed by Studio Libeskind, would be classed as requiring to meet the expectations of standard structural steel. Fabrication and erection by Walters Inc.

AESS 1 – BASIC ELEMENTS

This is the first step above standard structural steel. This type of application would be suitable for "basic" elements that require enhanced workmanship. Exposed structures of this kind could be found in roof trusses for arenas, warehouses, big-box retail stores and simple canopies and should only require a low-cost premium in the range of 20% to 60% due to its relatively large viewing distance as well as the lower-profile nature of the architectural spaces in which it is used.

AESS 1 characteristics include:

- The surface preparation of the steel must meet the standard of the Society for Protective Coatings SSPC-SP6.[1] This assumes commercial blast cleaning.
- All of the sharp edges are to be ground smooth.
- There should be a consistent weld appearance for all welds.
- It is assumed that bolted connections will use standard structural bolts (including tension control bolts, TC, if available in your region).
- Weld spatters, slivers and surface discontinuities are to be removed.

The characteristics will be explained in depth and illustrated in the following chapter (see page 43).

The Ricoh Coliseum in Toronto, ON, Canada, designed by Brisbin Brook Beynon Architects, uses trusses for its roof. A limited level of precision in fabrication would be required, given the building's use.

This steel truss system spans the central atrium of the Gaylord Texan Resort in Dallas, TX, USA, designed by Hnedak Bobo Group, Inc. Its distance to view allows for the use of slightly enhanced workmanship, predominantly standard steel sections and bolted connections. The structural system is laid out consistently. The more complex nature of the structure would incur a higher-cost premium.

Lighting must be taken into account when designing with AESS and selecting the appropriate category. The organization and patterning of the steel system of the Gaylord Texan Resort is more evident at night than during daytime. Sunlight and self-shadowing from the structural elements can mask details, whereas the blackness of the night sky in contrast with uplighting can make the structure stand out.

AESS 1 applications will see the use of fairly straightforward section types such as W (Universal), HSS, and often OWSJ and exposed profiled decking. Generally, this type of framing might appear similar to standard structural steel applications with the only difference that it is exposed to view. This fact means that more care is required to ensure that the standard structural members are aligned in a uniform way, that spacing is kept consistent and that the surfaces of the members are properly prepared to accept continuous finishes and coatings. A higher level of consistency in the use of connections, bolts and welds is also required.

The steel trusses supporting the skylight in the renovation to the Art Gallery of Ontario in Toronto, ON, Canada, designed by Frank Gehry, are AESS 1 type elements. Their technical detailing forms an intended contrast to the historic nature of the walls and the smooth curved wood surfaces of the feature stair.

The Semiahmoo Library in Surrey, BC, Canada, designed by Musson Cattell Mackey Partnership and Darrell J. Epp Architect Ltd., uses OWSJ and steel decking. This simple solution would correspond to the AESS 1 category with a very-low cost premium.

The exposed steel at the Dalian Conference Center in Dalian, China, is a good example of a very basic exposed structure that requires little enhancement or cost premium as the details are located well out of view. Castellated beams are relatively easy to manufacture and offer a simple enhancement to the aesthetics of the space.

The wider range of cost premiums associated with this category is largely a function of the nature of the section types and connection details. Projects using OWSJ and beams will have a lower cost than those employing trusses, as the fabrication costs associated with trusses are higher due to increased labor. Hollow structural sections will incur additional labor if fabricated into trusses, so it is more common to see HSS used as columns in this category. The cost premium will depend on the complexity of the design.

These types of applications may or may not require special fire-protective design. This is determined as a function of the use of the space and local laws. In some situations the steel may be left completely unprotected or sprinklered and will need to receive only a paint finish. Intumescent coatings could be used where the rating is to be one hour or greater, still permitting the exposure of the steel. The detailing on AESS 1 elements should not be greatly impacted by the relative thickness or finish of the intumescent coating, as much of this type of steel construction will be located well above eye level and out of range of touch.

As many AESS projects will specify more than one category of steel, it will be common to specify AESS 1 for the ceiling elements of a design where the distance to view is in the 6m/20ft or greater range, and specify a different class of AESS for those elements, like columns, that are located at closer proximity. In more basic applications AESS 1 can be used for the entire structure.

AESS 2 – FEATURE ELEMENTS with a View Distance > 6m/20ft

This category includes structures that are intended to be viewed at a distance larger than 6m/20ft. The process requires basically good fabrication practices with enhanced treatment of weld, connection and fabrication details as well as tighter tolerances for gaps and copes. This type of AESS structure might be found in retail and architectural applications where a low to moderate cost premium in the range of 40% to 100% over the cost of standard structural steel would be expected.

AESS 2 characteristics include all of the characteristics of AESS 1 plus the following features:

- Visual samples may or may not be required depending on the scope of the project.
- Standard fabrication tolerances of no more than one half as compared to standard structural steel constructions are required.
- Fabrication marks – number markings put on the members during the fabrication and erection process – should not be apparent.
- The welds should be uniform and smooth.

The characteristics will be explained in depth and illustrated in the following chapter (see page 46).

AESS 2 will generally be found in buildings where the expressed structure forms an important, integral part of the architectural design intent. The defining parameter of a viewing distance greater than 6m/20ft will infer that you might find this sort of steel in high-level roof or ceiling applications. For this reason AESS 2 steel may be specified for the distant components of a structure, and higher-quality AESS for the low-level elements of the structure. These should be clearly marked on the drawing sets so that the treatments can be differentiated and the respective cost premiums separated out.

The trusses that span the future CIBC Pan Am / Parapan Am Aquatics Centre and Field House in Toronto, ON, Canada, designed by NORR Ltd., are fabricated to AESS 2 standards. They use a combination of W (Universal) and L-shapes as well as welded and bolted connections. The column supports were fabricated to AESS 3 standards. Fabrication and erection by Walters Inc.

It will be more common to see W or HSS members specified for this category rather than industrial-looking members such as OWSJ. This type of application may use a combination of bolted and welded connections. As the viewing distance is great there is normally less concern about concealing the connection aspects of larger pieces to each other. Therefore splice connections will be designed to be discreet but not imperceptible. There is no grinding of welds in this category.

The International Terminal at Calgary International Airport in Calgary, AB, Canada, designed by DIALOG, uses AESS 2 standards for the trusses that span the hall. They are fabricated from a combination of round and square HSS. The major lengths of the trusses are achieved using fully welded connections with no weld remediation. The columns and struts are AESS 3 due to their proximity and specific connection detailing.

A closer view of the truss shows that a variety of sizes of HSS have been used, dimensioned to suit the loading as well as to allow for ease of welded connections between members. Pin, hidden and discreet connections have been used to join the segments of the truss. Fabrication and erection by Supermétal.

The cost premium for AESS 2 steel ranges from 40% to 100%. There may be lower costs associated with the clean use of standard structural shapes with bolted or simple welded connections and higher costs associated with the use of HSS shapes, complex geometries and a predominance of welded connections. As one of the common applications of AESS 2 members will be for roof, skylight or ceiling support systems, the fire protection method must be defined from the outset of the project. If intumescent coatings are used, these can help to conceal any inconsistencies in surface conditions. Given the feature nature of this steel, greater care will also be required in the coordination of mechanical systems that will be integrated into ceiling and roof systems.

The triangular arched truss that supports the fabric roof of the Bank of America Pavilion in Boston, MA, USA, designed by A.Form Architecture pc, would be classed as AESS 2. It makes use of unremediated welded connections for the tube-to-tube connections within the truss. The truss segments are bolted together. The curved shape is achieved through the use of straight segments, simplifying fabrication. This works as a factor of the scale of the building.

AESS 3 – FEATURE ELEMENTS
with a View Distance ≤ 6m/20ft

This category includes structures that will be viewed at a distance of up to 6m/20ft. This is suitable for feature elements where the designer is comfortable allowing the viewer to see the art of metalworking. The welds should be generally smooth but visible and some grind marks would be acceptable. Tolerances must be tighter than normal standards. As this structure is normally viewed closer than 6m/20ft it might also frequently be subject to touch by the public, therefore requiring a smoother, more uniform and more durable finish and appearance. This type of structure can be found in airports, shopping centers, hospitals or lobbies and is expected to incur a moderate cost premium that could range from 60% to 150% over standard structural steel, as a function of the complexity and level of final finish desired.

AESS 3 characteristics include all of the characteristics of AESS 1 and 2 and also the following features:

- The mill marks are to be removed so as not to be visible in the finished product. Removal of these marks would typically be accomplished by grinding.
- Butt and plug welds are to be ground smooth and filled to create a smooth surface finish.
- The weld seam intrinsic to HSS shapes is to be oriented to reduce its visibility.
- Cross-sectional abutting surfaces are to be aligned.
- Joint gap tolerances are to be minimized.
- AESS 3 feature elements may require all-welded connections. This is noted as optional among the typical features, acknowledging that a particular aesthetic might purposefully choose bolted connections.

The characteristics will be explained in depth and illustrated in the following chapter (see page 50).

When AESS elements are brought into close range for view as well as potentially for touch it is necessary for the team to come to a clear understanding about the level of finish that is either required or expected of the steel. The natural look of welds that would be out of view in the AESS 2 category will now be visible to those present in the space. Simple bolted connections may need to be designed to look more artful if they are to become part of the design language. Connections will come under closer scrutiny so their design, tolerances and consistent appearance will become more important. The workmanship required to improve these beyond both standard structural steel and AESS 1 and 2 levels may have a significant impact on the cost of the overall structure. If the required attributes or characteristics of the steel are not thoughtfully considered, the steel for the project can easily be priced out of viability.

AESS 3 allows for grinding to remediate or mask welds. This is not to infer that grinding is required or even necessary, but just that it can be considered. It is advisable to discuss this with the fabricator as details are being designed. There may be other, less expensive options that work equally well.

The approach to steel detailing at the Palais de Congrès in Montreal, QC, Canada, designed by Les architectes Tétreault, Dubuc, Saia et associés with Hal Ingberg architecte, combines the technical and the smooth. Where the larger splices in the truss employ accentuated plates with bolts, the crossing of the diagonal W (Universal) section through the bottom chord of the truss has been ground smooth. An intumescent fire-protective coating masks some of the workmanship.

The cutout detail on the Palais de Congrès also complies with AESS 3 standards. The extra care required to cut and weld the curved plate detail inside of the cutout adds to the cost of the project.

The cost premium to be encountered in AESS 3 steel will depend greatly upon the types of members chosen, the nature of the connections, and the desire of the designer to either conceal or express the materiality of the steel itself. It is assumed that some effort will be put into further preparation of the surface to increase its smoothness in order to ensure that some of the natural finish on the steel and mill marks do not show through the paint.

The Chihuly Pavilion in Seattle, WA, USA, designed by Owen Richards Architects, has different requirements on the steel due to the continuity of the members from close to distant view. AESS 3 would be appropriate to govern the choice with deference to close view. Some savings can be made in the design of the high-level connections that are out of view range. The added challenges of working with curved steel will be more suited to this category and also increase costs.

When welds cannot be done in the shop, where conditions are more controlled and jigs can be used to ensure precise alignment of the components, it must be realized that a large amount of site welding of complex elements will result in cost premiums. Some site welds may not be of the same quality as can be expected of shop welds. Welds are expected to be of a higher quality than those for AESS 2 structures, where the welds would be out of view and touch due to their height. AESS 3 welds are expected to have a very uniform appearance. Although some touch-up grinding of the welds may be required to ensure uniformity, complete grinding of all welds would not be included in the features of this category of steel. It is assumed that good-quality, uniform welds would be left exposed.

The pedestrian bridge at the Museum of Flight in Seattle, WA, USA, designed by SRG Partnership Inc. and Magnusson Klemencic Associates, is served by AESS 3 type steel due to the close view maintained even to the height of the bridge. The added complexity of the curvature and welding of the round HSS "hoops" adds to the cost of this structure.

Where bolted connections are employed more care will be taken to ensure that there is an aesthetically based consistency in the connections that will likely require more fabrication time and potentially more material. Simple approaches such as ensuring that all bolt heads are located on uniform sides of the connections can greatly enhance the details with little extra cost. If bolted connections are required for erection ease but are visually unacceptable, concealed connections can be employed to give the appearance of a seamless or welded connection without the associated price tag. For these types of connections the attaching plates are kept within the general line of the members, so that cover plates can be attached over the bolted elements. Concealed connections must be made weatherproof for exterior applications (for more information on hidden connections see page 140).

All in all, the underlying idea of the AESS 3 category is to change the appearance of the final product to make it smoother to the eye, recognizing that it is not always necessary to use more expensive fabrication techniques to achieve this effect.

The Arganzuela Bridge in Madrid, Spain, designed by Dominique Perrault, is able to use the scale of the structure to keep the detailing of the welded connections of the curved box sections to AESS 3 standards. The connections within the members use either unremediated welds or bolts.

AESS 4 – SHOWCASE ELEMENTS

Showcase or "dominant" elements should be used where the designer intends that the form is the only feature showing in an element. All welds are ground and filled, edges are ground square and true. All surfaces are sanded and filled. Allowable tolerances of these fabricated forms are more stringent, generally to half of the standard tolerance for standard structural steel. All surfaces are to be "glove-smooth". The cost premium of these elements can range from 100% to 250% over the cost of standard structural steel – completely as a function of the nature of the details, the complexity of construction and the selected finishes.

The characteristics for AESS 4 include all previous and in addition the following features:

- The normal weld seam in an HSS member should not be apparent.
- Welds are to be contoured and blended.
- Steel surfaces are to be filled and sanded.
- Weld show-through must be minimized.

The characteristics will be explained in depth and illustrated in the following chapter (see page 54).

AESS 4 Showcase Elements represents the highest standard quality expectations on AESS. The products and applications are diverse and there is a wide variety of member types. Many of the column or spanning members are custom-fabricated. In some cases this may be due to the very large size and structural capacity required of the member. In other cases it is due to the particular architectural style desired in the exposed structure. Many of the members tend to employ steel plate that has been custom-cut to odd geometries. Such geometries, when not based on a combination of simple "circular holes" and straight cuts, will greatly increase the fabrication costs of the project.

Beijing Capital International Airport in Beijing, China, designed by Foster + Partners with Arup, could be considered as AESS 4 in its widespread use of all-welded connections for the curved round HSS trusses that brace its front wall. The semi-gloss finish on the trusses requires excellent surface conditions and the removal of fabrication marks. The use of curved steel adds to the complexity and cost of the structure.

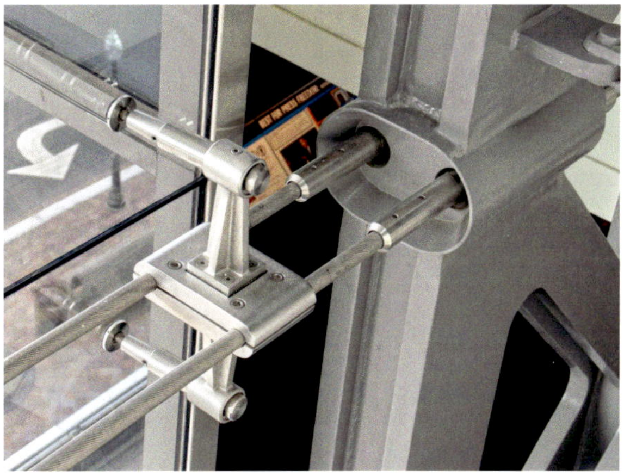

The Newseum in Washington, DC, USA, designed by Ennead Architects with Leslie E. Robertson Associates, makes fairly exclusively uses specialty plate steel for the vertical trusses. The sharp edges and clean lines are characteristically AESS 4. Crisp edges require more care in the application of coatings to ensure good multi-coat coverage at the corners. This truss is located away from danger of chipping due to pedestrian traffic. The extra care and expense in this level of fabrication is warranted as the structure is visible at very close range. It also must coordinate its tolerances with cable and spider glazing systems.

On many of these projects the edges of the steel have been finished to be very crisp and precise. The straightness of the lines of these members is a critical aspect of their fabrication.

The pedestrian Peace Bridge in Calgary, AB, Canada, designed by Santiago Calatrava, has used a high degree of brake forming, welding, grinding and filling to create this tubular shape. A structure like this would sit at the extreme high end of AESS 4 expectations and cost premium.

AESS 4 makes extensive use of welding for its connections. In most cases the weld is ground smooth and any member-to-member transitions are filled and made extremely seamless in appearance. This care is applied to the fabrication of on-site connections and expenses are incurred to carry out on-site welding, grinding and filling. AESS 4 on-site connections will tend not to make use of simpler discreet, hidden or bolted connections for reasons of aesthetics.

King's Cross Station in London, England, designed by John McAslan and Partners with Arup, incorporates solid custom castings. The use of large custom castings will generally call for AESS 4 standards due to the added challenges in design, fabrication, remediation and erection.

The World Financial Center entry pavilion uses custom-bent steel to create this pair of diagrid column supports. The connections are all welded, incurring additional expense and challenges on the site to prepare scaffolding to carry out the work. The expectation was that the numerous on-site welds would be remediated to the same high quality as those done in the shop.

Special members often require additional care in transportation and handling as the maximum amount of work, including painting, is typically carried out in the fabrication shop to maintain the highest quality of work.

The important issue is to note what triggers the shift from AESS 3 to AESS 4 standards as it might seem quite subtle. Although grinding is an option at the AESS 3 level, many projects tend to avoid it. Remediation is quite extensive at the AESS 4 level. While characteristics will be discussed more in later chapters, here are some emerging tendencies that would be included in AESS 4 detailing:

- mostly custom sections, many created from plate material
- significant remediation of welding throughout the elements
- curved steel – although this may also be used in lower-category projects, it does add challenges
- use of large custom castings
- sharp corners on members
- seamless appearance
- absence of W (Universal) shapes
- splices between transportable sections done via welding (as opposed to bolting, hidden or discreet connections)
- often coordinated with cable-supported glazing systems, necessitating tighter tolerances

AESS C – CUSTOM ELEMENTS

What is critical to the establishment of the AESS Category System is to ensure that the AESS 1 through 4 categories are used without altering their associated characteristics. By contrast, if a project or elements of a project require alterations to the characteristics assigned to the standard categories, then Custom AESS is designed to select the characteristics in an "à la carte" fashion.

There are certain applications of AESS that will require a customized approach to design and specification. With increases in the reuse of steel for sustainably-minded projects, unique sets of criteria will come into play. Requirements will address the presence of finishes, corrosion, inconsistencies between members and whether a project desires to showcase the reuse or else blend the extant material with new material. As some historic steel is fastened with rivets, different treatment may be required where new connections are mixed with old in order to create coherence.

The restoration and structural upgrade that is being carried out on Paddington Station in London, England, uses bolted connections in an effort to blend in with the existing riveted connections, which are no longer used.

Custom AESS elements will also include a use of steel that is more sculptural in nature. In some instances, the nature of the steel is intended to be a highlight of the project, while in other cases the nature of the steel is to be concealed and the product looks more plastic in nature. The latter may require a higher degree of finish and workmanship than would be expected even for structures in the AESS 4 range.

The Ram's Horn by John McEwen is a wind mitigation device located between high-rise towers in downtown Calgary, AB, Canada. The inspiration for the artist's piece was with an actual ram's horn and steel fabricator Walters Inc. replicated the "feel and texture of the horn in steel". The characteristic plates and the evidence of welding to the round HSS section that joins the plates through the center were deliberate in order to add material texture to the piece.

This detail of the Ram's Horn attests to the appreciation of the artist of the art of metal working. Steel is steel and welds are welds and in this custom work the tactile nature of the evidence of the fabrication process is deemed to add to the piece.

The AESS used in Heathrow Airport Terminal 5 in London, England, designed by Rogers Stirk Harbour + Partners, Arup and Watson Steel Structures Limited, is high-end AESS 4 with added requirements due to the use of custom castings, plate steel, sharp edges, prefabrication, extreme size, erection challenges and coordination with a multi-storey glazing support system.

The AESS requirements for the tree-like support system at the University of Guelph Science Building in Guelph, ON, Canada, designed by Young + Wright Architects with Walters Inc., are higher than AESS 4 due to the combined use of custom castings, mechanical pipe and a high-gloss paint finish. In this instance the Custom AESS requirements are based on AESS 4 and additional notes added to ensure the final high-quality outcome of the product. This would include greater than normal care in handling as well as the remediation of the orange-peel like surface inherent on the casting to make the finish more like that of the pipe.

The use of stainless structural steel is included in this category as this material has different specifications and particular issues that must be addressed to ensure a high quality of installation. Where a painted structure can be touched up after erection damage, increased care in erection must follow with stainless steel as damage or surface scratching is less easy to repair. Stainless steel also requires the use of dedicated equipment. Any tools used to fabricate, cut or polish regular carbon steel must never be used on stainless steel as they will embed particles of regular carbon steel in the stainless material and cause rust! Stainless steel has different structural capacities than normal carbon steel and these must also be addressed in the design of stainless steel structures. This material is highly recommended for continuous outside exposure due to its inherent corrosion resistance.

The replacement domes at Robson Square in Vancouver, BC, Canada, fabricated by George Third & Son, used stainless steel to create a more durable and weather-resistant solution for this outside structure. Extra care is required to ensure high-quality welding as typical remediation via grinding does not work with stainless steel finishes. Grinding can mar the surfaces and cause more problems, particularly when using round, tubular members.

The use of galvanized steel for this highly complex structure at Federation Square in Melbourne, Australia, designed by Lab Architecture Studio with Bates Smart Architects, asks for a Custom AESS set of requirements. The difficult geometries will increase the fabrication and erection time and cost, although the very technical look of the steel allows for an increased use of bolted connections. However, the design of these bolted connections is quite different from standard fare, with tighter tolerances required by the slotted fit as well as the exposed Y-shaped welds at member-to-member joints.

MIXED CATEGORIES

Mixed categories are to be expected on almost all projects. It will be very common to specify lower-level categories for roof/ceiling framing elements and higher-level categories for columns and sections that are nearer to view and touch. Mixed categories can be addressed by making notes on the engineering drawings and/or creating a schedule of members, types and categories within the specification. A maximum of two types is to be expected for simplicity and clarity. It is typical to use adjacent categories – for example AESS 2 for the trusses and AESS 3 for the columns – as this acknowledges the distance factor.

It is also possible to mix categories on individual elements. A case in point are sections that have a side exposed to view/touch and a side that is "buried" or otherwise hidden from view. Specifying different AESS levels in such a case is of great financial benefit when finishing extremely large members. There should be a special note associated with the custom member and the specific combination of categories noted. When using the AESS categories to this level of detail it is also advantageous to be sure that this is clearly and personally communicated to the fabricator prior to bidding the job. The fabricator may have some useful cost-saving suggestions that can positively impact the overall project.

The large triangular members that are part of the diagrid system for the Bow Encana Tower in Calgary, AB, Canada, designed by Foster + Partners with Zeidler Partnership and fabricated/erected by Walters Inc., have very sharp edges on the front face that are designed to AESS 4 precision. As the rear face of the member will be embedded this is not designed to AESS specifications.

The rear of the large diagrid nodes for the Bow Encana Tower are finished as standard structural steel, in contrast to the AESS 4 finish of the face that is exposed to the atrium. The rear side is covered by the curtain wall cap system and will not be exposed to view. There is great cost saving in differentiating the AESS approach to the two sides of the member.

PORTO INTERNATIONAL AIRPORT | PORTO, PORTUGAL

The AESS used in Porto Francisco Sá Carneiro International Airport in Porto, Portugal, designed by ICQ with WS Atkins, has clearly taken different design approaches for the steel trusses that span across the departure hall vis-à-vis the exposed steel that frames the roof around the skylights.

A close view of connections on the bottom chord of the large triangular truss shows the detailing of the tension-to-tube connections. There is much precision required to fit the components. The welds are neatly done and left exposed, which could place this element at the high end of AESS 3 for cost. The fitting required for trusses with such tension systems will increase costs.

The skylight framing is more finely detailed as it includes a specialized tension system to support the glazing. There is a clear contrast between the types of members chosen for the skylight glazing support system and the adjacent roof framing system that requires a different specification for a lower category of AESS. Note the use of W members up high and tubular members at the lower view levels.

CHARACTERISTICS

CHAPTER 3

The Arganzuela Bridge in Madrid, Spain, designed by Dominique Perrault, uses a variety of custom fabrication methods.

A set of characteristics is associated with each category as outlined in the previous chapter. Higher-level categories include all of the characteristics of the preceding categories, plus a set of specific additional requirements.

AESS 1 – BASIC ELEMENTS

AESS 1 – Basic Elements is the first step above standard structural steel.

1.1 The surface of the steel must be prepared using the standard of the Society for Protective Coatings for commercial blast cleaning (SSPC–SP6).[1] Prior to blast cleaning, any deposits of grease or oil are to be removed by solvent cleaning. Commercial blast cleaning is intended to remove all visible oil, grease, dust, mill scale, rust, paint, oxides, corrosion products and other foreign matter, except for spots and discolorations that are part of the natural steel material. As a consequence, there should be no problems with the range of finishes that would be required for AESS 1 type applications.

Steel prior to commercial blast cleaning (left) exhibits natural oxidation and may have other materials on the surface that would prohibit proper adhesion of finishes. After commercial blast cleaning (right), the steel will oxidize prior to final finishing, but it is now more effective to remove problematic residues prior to painting and finishing.

Surface preparation standards commonly used for structural steel elements only provide for power tool-type cleaning and should not be relied upon for consistent finishes in AESS applications. The elements to be cleaned need to fit in the blast chamber; elements may include not only base material but also partially assembled pieces. Commercial blast cleaning is required prior to the application of any primers. At areas where welding is to occur, primers will be held back and oxidation needs to be removed again prior to welding. Removal of recurrent oxidation is done with hand tools.

1.2 All sharp edges are to be ground smooth. Rough surfaces are to be deburred and ground smooth. Sharp edges, resulting from flame cutting, grinding and especially shearing are characteristic of plain structural steel and considered unacceptable in any AESS application as they do not lend themselves to any sense of precision in the final fabrication and installation.

Cut steel is naturally rough. As a basic requirement for AESS applications, sharp edges and corners must be ground smooth.

1.3 There should be a consistent appearance for all welds. Intermittent welds can be made to look continuous, either with additional welding, caulking or by using body filler. For corrosive environments, all joints should be seal-welded. The natural seams of hollow structural sections are acceptable as produced.

Intermittent welds are structurally sufficient for this connection. The use of a filler to create the appearance of a continuous weld saves costs and does not compromise the performance of the welded connection.

Fabricators are often asked to create continuous welds even when these are structurally unnecessary. This adds extra cost to the project and takes additional time. If not structurally required, the welds in themselves need not be continuous. Prior to the application of the final finish, appropriate caulking or filler can be applied between the intermittent welds to optimize the appearance. Filling also helps in the cleaner application of finishes and prevents the build-up of dirt in the joints. Care must be taken in the application of filling materials to ensure that the surfaces are clean to achieve adherence and that the compounds are compatible with the finish.

Tubular sections have a weld seam along their length due to their manufacturing process. In AESS 1 applications these need not be remediated. The larger the diameter of the member and the thicker the steel wall, the larger the seam.

The feature stair at the Boston Society of Architects in Boston, MA, USA, designed by Höweler + Yoon Architecture, shows a very artful use of the intermittent weld as a feature of the detailing of a connection. Although the stair is constructed at a higher AESS category, and such coordination of the location of the welds with the bolts would qualify as a Custom AESS category, it does allow us to question the need for a continuous weld appearance in all cases. The detail also highlights an aesthetically acceptable use of standard structural bolts.

In some cases, particularly when joining very long slender members, the heat applied to create a continuous weld can cause the member to warp and must therefore be avoided. Filled intermittent welds provide the better solution.

1.4 The bolt head orientation may be specified. It is assumed that bolted connections use standard structural bolts. The heads of the bolts should all be located on one side of the connection, but they cannot be fastidiously aligned. There should also be consistency from connection to connection. In any case, the structural tightening of the bolts must take priority.

Specifying to AESS 1 would negate issues arising from inconsistency of the "side" of the head as illustrated in this detail. Such a simple, cost-effective requirement can greatly improve the appearance of the connection.

Tension Control (TC) bolts may be used instead of regular Hex Head bolts (top). Tension Control bolts only require access from the threaded end for tightening.

Within AESS 1 (bottom), better control of bolting details would be expected.

The process of welding results in spatters. One of the benefits of shop welding is easier access for the ironworkers. This applies both to the welding process and to any remediation requirements.

1.5 Weld spatters, slivers and surface discontinuities are to be removed as these would mar the surface and likely show through the final coating. Weld projection of up to 2mm / .08in is acceptable for butt and plug-welded joints. This applies to procedures carried out in the fabrication shop prior to erection as well as to weld spatter and surface discontinuities that may occur during or as a result of erection. When temporary steel supports or shoring elements are removed, the marred surfaces and any rustproofing or coatings should be properly repaired prior to final finish applications. In order to avoid potential conflict about the admissible minimum diameter or intensity of spatter, all weld spatter is to be removed.

AESS 2 – FEATURE ELEMENTS with a View Distance > 6m/20ft

AESS 2 – Feature Elements applies to structures intended to be viewed at a distance > 6m/20ft. AESS 2 characteristics include all of the characteristics of AESS 1 plus the following features:

2.1 Visual samples may or may not be required depending on the scope of the project. They are considered optional with view to issues of suitability, cost and scope. Not all projects warrant the level of certainty that is provided by actually seeing the final product. If working with a known fabricator, an inspection of existing work can provide a basis of discussion as to the aims of the detailing. Depending of the availability of such references, visual samples can range from small pieces of fabrication that may include connections or finishes, to full-scale components. Visual samples used to validate the intention of the final installed product can be:

- 3D rendering
- physical samples of surface preparation and welds
- a first-off inspection of the first element fabricated for use in the finished structure
- scaled or full-scale mock-ups. Mock-ups are to demonstrate aesthetic effects as well as qualities of materials and execution. They may display the finished surface. The architect's approval will be required and the mock-ups retained until the project is completed. Approved full-scale mock-ups may be integrated as part of the final project.

In some cases an agreement to incorporate full-scale mock-ups in the final project can make practical and economic sense. It is very important to bear in mind the potential for delay and additional costs by the requirement of physical visual samples in the timeline of the project. If a fabricator is expected to create a large element, this will delay the fabrication of similar elements until the approval is reached. The particular requirements for each project or particular special assembly should be specified in the contract documents.

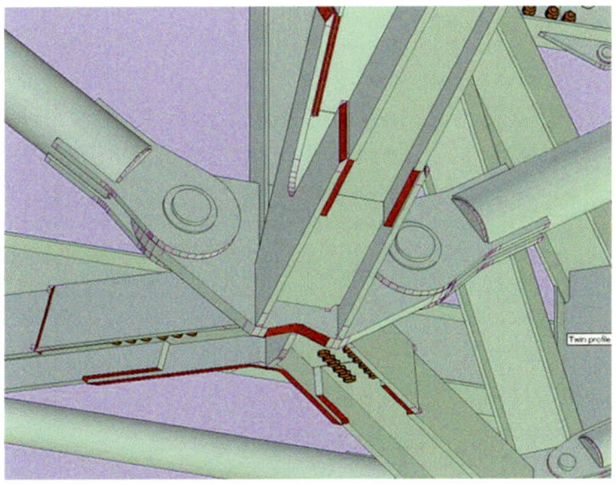

For the Ottawa International Airport, the steel fabricator Walters Inc. provided a detail from their Tekla Structures software to Brisbin Brook Beynon Architects for approval. It shows quite clearly the nature of the connection, platework and types of welds.

In comparison with the actual completed detail, the rendering quite accurately depicted the detail.

The fabricator of the World Financial Center entry pavilion, Walters Inc., prepared this sample for the tube to plate ring condition of the diagrid columns to allow the architect to understand how three different welded joints would look. This could be related to cost premiums on the project.

A full-scale mock-up was fabricated by Walters Inc. to show the geometry and finishes for Pearson International Airport in Toronto, ON, Canada, designed by SOM. This very large element was incorporated into the final project, in spite of some small modifications being made to the rest of the elements.

The slightly different mock-up is impossible to locate in the finished project.

As large physical mock-ups are normally viewed at the fabricator's shop, some very important realities are to be kept in mind – lighting, distance and orientation. The lighting at the shop will be quite different from the eventual site condition. If the element to be inspected has an intended distance of view, it should not be inspected at close range. Also the angle of view or stance may be quite different from the final piece. All these conditions may affect the perception of the element. Sometimes the element can be lifted or turned to more closely approximate the final view.

During the shop visit to inspect the mock-ups for the Calgary International Airport in Calgary, AB, Canada the fabricator Supermétal was able to lift this strut into a position similar to its final use position to better allow a distant view.

2.2 Standard fabrication tolerances of no more than one half as compared to standard structural steel constructions are required. This is to recognize the increased importance of "fit" when assembling these more complex components.

Large tolerances can lead to a sloppier appearance and lack of uniformity in the connections and potentially to problems in the erection of complex geometries. This has a direct impact on the erection process and the potential cost implications of making site modifications to members that do not fit. The level of "fit" is essential for all structural members, plates, angles and components comprised in a project. In highly articulated projects cumulative dimensional error can be disastrous in the fitting of the final elements of each erection sequence.

2.3 Fabrication marks (number markings put on the members during the fabrication and erection process) should not be apparent. Fabrication marks can include punch marks that go into the steel, written notes put on the steel and stickers with bar or QR codes. In some instances punch marks can be left but located away from view. In other cases they may be lightly ground out. They could also be filled prior to finishing. The treatment may vary throughout the project by member and location. Surface markings may be concealed by coatings or they may require light sanding. Stickers and any residue should be removed to prevent issues with the application of finishes.

The fabrication marks on the steel have a yellow box around them. These are used to track the material in the shop. As they are created by punching numbers into the steel, they can be filled and sanded to make them less visible.

2.4 The welds should be uniform and smooth. Quality welding is essential in AESS 2 categories and higher. Good-quality welds can save substantial cost in a project. If welds are uniform and consistent in appearance there may be less need for post-treating. Too many welded connections are subjected to needless grinding. Welding is a natural condition of steel connections, and if neatly done, should be allowed to remain as part of the final product.

This all-welded truss in one of the Dubai Metro Stations in Dubai, UAE, designed by Aedas Architects, allows the welds to be simply exposed. These would have been shop-fabricated, allowing for better uniformity.

The intention to have more of the welds carried out in the fabrication shop can impact the design of joints as well as the transportation of potentially larger pre-assemblies and the erection on site. This does not infer that high-quality site welding is not possible, only that it might incur a cost premium over shop welding.

AESS 3 – FEATURE ELEMENTS with a View Distance ≤ 6m/20ft

AESS 3 – Feature Elements applies to structures intended to be viewed at a distance ≤ 6m/20ft. This increased proximity begins to place the evidence of certain fabrication processes into close range where the final product can be both viewed and touched. In many cases these markings will either need to be carefully positioned so that they cannot be seen, or else removed or concealed.

3.1 Mill marks are to be removed so as not to be visible in the finished product. Removal of these marks would typically be accomplished by grinding.

Mill marks are rolled into the steel and protrude from the surface. To remove a mill mark it must be ground, sanded and potentially filled before painting.

Some mill marks may just be turned away from view, but this must be carefully considered and discussed with the fabricator/erector as the location of the marks is difficult to control.

3.2 Butt and plug welds are to be ground smooth and filled to create a smooth surface finish, caulking or body filler are acceptable. Butt welds are very commonly used when splicing plate together to create larger entities. In many instances the appearance of the splice is not desired and so remediation is necessary to create a seamless appearance. These kinds of welds can bring along additional material or slight depressions in the members. Such imperfections would be visible after finishing. If additional material is present, it should be ground smooth. If there are depressions, the voids can be filled with body filler and the surface ground smooth. This is more challenging on curved surfaces due to the geometry, while flat surfaces are easier to work as the reference plane is consistent.

The Arganzuela Bridge (left) makes widespread use of natural butt welds where the plates have been joined. The technical nature of the bridge design as well as the dappled lighting help to make this condition recede from view. These unremediated butt welds would then be classed as AESS 2, even though the complexity of the structure might suggest a higher category.

The butt welds on these plates (top) have been ground smooth. A smooth result is easier to obtain on flat material, prior to its incorporation into an assembly. Grinding is more difficult on complex geometries.

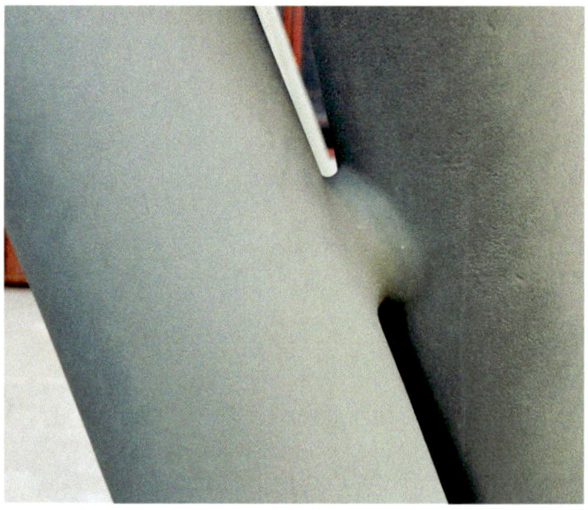

In this detail on the World Financial Center entry pavilion, solid round bars have been used to connect pairs of hollow round sections. After the weld is done (left) it is ground smooth (middle) to remove any weld material that would sit proud of the finished surface. It is then filled with body filler and sanded to make the connection disappear. The final connection (right) reveals the high level of finish expectations on this project. The degree of filling and contouring of the intermediate connector would class this as AESS 4.

3.3 The weld seam intrinsic to HSS shapes is to be oriented for reduced visibility. All seams are to be oriented away from view in a consistent manner, or as indicated in the contract documents.

This detail could have been improved by orienting the weld seam on the left-hand HSS member to the rear. This has been done for the member to the right of the column.

On structures that use HSS, such as the Caisse de Dépôt et Placement in Montreal, QC, Canada, it is requested that all of the seams are oriented away from view. This measure can be quite effective if the truss is directional so that the seams can be oriented towards the glazing behind – more effective than to have them ground, sanded and filled. Lighting, both natural and artificial, can either highlight or mask the appearance of the seams.

Welded seams are part of the manufacturing process of HSS members and a natural finish appearance. A seamless finish is not possible without significant additional expense and time. There are alternatives to grinding the seams. They can be consistently located to give a uniform appearance. If HSS seams are able to be oriented away from direct view, this is an acceptable solution. If the seams are located in members whose viewing angles are multiple, greater care must be taken in detailing the members to achieve a consistent look. If two HSS members are joined, then ensure that the weld seams are aligned.

3.4 Cross-sectional abutting surfaces are to be aligned. Ensuring that the steel conforms to half standard tolerances (see above, 2.2) will not guarantee completely precise alignment of abutting members – particularly when using off-the-shelf structural sections that will have had little specialty fabrication work done to them. Surfaces may have to be shaped or ground at the point of connection to ensure that they are aligned. In some lighting conditions shadow casting may be problematic. Where the inconsistencies are small, advanced knowledge of the final finish coat may help to conceal such slight misalignments.

Even this simple bolted connection in a transit station (top) is made to look better by taking care that the abutting surfaces align.

Even though the connections are bolted at the Ronald Reagan National Airport in Washington, DC, USA, designed by Pelli Clarke Pelli Architects (right), the very tight and uniform spacing between the adjoining members provides a crisp clean look. Although tight tolerances like these make the steel more difficult to erect, this is less expensive than an all-welded option.

3.5 Joint gap tolerances are to be minimized. This characteristic is the equivalent to 2.2 above. A clear distance between abutting members of 3mm/.12in is required. In keeping with tighter tolerances on the members, the reduction of joint gaps in bolted connections, which are quite common in many AESS applications, aids in ensuring consistency and tighter design.

3.6 AESS 3 Feature Elements may require all-welded connections. This is noted as optional, acknowledging that a particular aesthetic might purposefully choose bolted connections.

This truss has used fully welded connections. The diagonal tubes have been sized smaller than the four corner tubes to allow for space to more easily blend the welded connection. It is located at Waiward Steel's fabrication shop in Edmonton, AB, Canada, and is designed to showcase their level of workmanship.

While there are instances where a welded connection is required anyway, many connections are more easily created with bolts. This is particularly true on site, as welded connections are easier to fabricate in the shop. It is critical to discuss among the team, including the fabricator, how to best handle the desire for welded connections, as there are other options for making the necessary splices between sections. These are referred to as "discreet or hidden connections" and will be addressed in *Chapter 8: Specialized Connections.*

Welded connections can add cost to a project. The erection conditions might require temporary shoring to hold the geometry in place for on-site welding. There may be additional work to repair the surfaces that have been damaged due to the removal of the temporary steel. In some situations, whether due to access constraints or issues of time, welded connections might not be possible at all. Alternatively, if an all-welded appearance is desired, hidden bolts may be considered as an acceptable solution. If a hidden connection concealed behind a cover plate is used in an exterior environment, care must be taken to seal the joint to prevent water from becoming trapped.

AESS 4 – SHOWCASE ELEMENTS

AESS 4 – Showcase Elements or "dominant" elements have as their design intention that the form is the only visible feature, while the technical nature of the steel is to be hidden or downplayed. All welds are ground and filled; edges are ground square and true; surfaces are sanded and filled to be "glove-smooth". Tolerances are generally to half of the standard tolerance for structural steel.

4.1 The normal weld seam in an HSS member should not be apparent. If it is not possible to orient the weld seam away from primary view, or if the viewing angles to the structure are from all sides, the seams may need to be ground and filled. In some instances where there are numerous weld seams to conceal it might be more prudent to choose mechanical pipe over round HSS. Mechanical pipe has the advantage of being seamless; however, it also has different physical properties so that alternate approaches may be required in cases where fabricating details, like workability, are quite different from standard hollow structural sections.

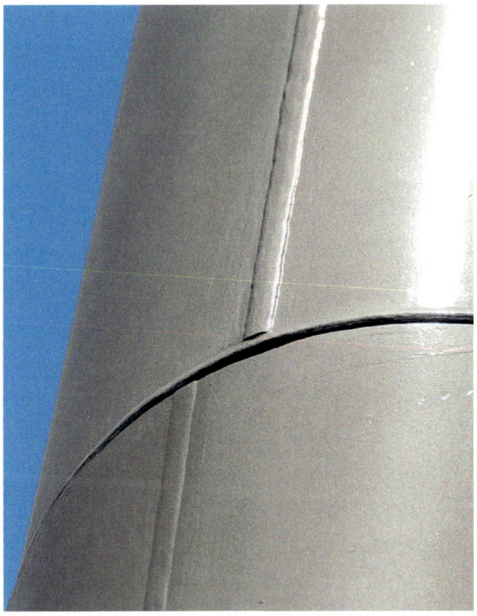

As HSS members increase in diameter and thickness, so does the seam. There are many inexpensive means to either conceal this seam or incorporate it into the design. This detail shows a clear lack of communication in the project and the need for the treatment of seams to be acknowledged. At minimum these seams should line up. This detail is fairly near to view and quite noticeable.

The round tubular members at the Rose Center for Earth and Space in New York, NY, USA, designed by Ennead Architects, use mechanical pipe instead of round HSS. This obviated the need to deal with the weld seams, which would have been difficult to either remediate or orient as the structure has multiple view angles.

4.2 Welds are to be contoured and blended. In pursuit of an overall contoured and blended appearance, welded transitions between members are also required to be contoured and blended.

Where large all-welded elements such as trusses are unable to be shipped in one piece, on-site weld splicing may be required. It is extremely important to understand the process by which such finely finished splice connections are made and remediated, as it will add cost to the project in terms of time and money. These sorts of splices should be considered exceptional and not the rule, being reserved for connections in AESS 3 or 4 type members that are close to view and that warrant such treatment.

An extremely fine appearance was desired for the exposed steel trusses at the 442m/1,449ft tall KK100 Tower in Shenzhen, China, designed by TFP Architects. Here the weld appearance has been minimized through grinding, sanding and filling. This truss is viewable at a very close distance.

FABRICATION OF SEAMLESS TUBE-TO-TUBE SPLICE CONNECTIONS

The process starts with male and female members that have temporary side tabs attached to allow for temporary bolting during welding. After welding the tabs are cut off with a torch and the weld and tab removal marks are ground away. The primer has been held back from the welded connection as it would interfere with the process. A body filler is applied to the surface to fill any marks from the grinding process. The surface is then sanded and will be primed prior to the element receiving its final paint finish.

This welded tube-to-tube detail at the Rose Center for Earth and Space has the welds ground and filled, creating an extremely smooth transition. This detail is located near eye level and truly required such remediation to maintain the sculptural feel of the design.

This type of detailing should be reserved for the most particular applications, those in very close proximity for view and touch and those whose form, fit and finish require a seamless appearance. Grinding and contouring welds is time-consuming and thereby very expensive. It is more easily done in the fabrication shop, in a controlled environment and where the pieces can be manipulated (by crane if necessary) so that the ironworkers can properly access the details. In-situ high-quality welding might require the erection of additional secure platforms to access the welded connections. Therefore, negotiation for this type of detailing must look at maximizing the sizes of the pieces to allow for shop fabrication, and minimizing on-site work. This brings in issues of transportation and site access if the resultant members are very large. Also, such pieces must be carefully handled and stored on the site to prevent damage.

4.3 Steel surfaces are to be filled and sanded. Filling and sanding is intended to remove or cover any surface imperfections and can incur a high cost premium – a particular case in point that not all AESS needs to be created equal. Procedures such as this are not required if using a thicker intumescent finish. Great care must be taken to ensure that the filled and sanded surface is consistent with the finished surface of the adjacent steel, otherwise variations will be revealed after the finished coating is applied. Steel castings, for instance, have a different surface than adjacent HSS sections, so any surface treatment to the joining weld must mediate the two finishes.

Steel has slight surface imperfections and roughness as a result of its manufacturing process. While this might be acceptable in AESS 1 steel (as discussed above), it must be remediated for AESS 4 material.

The tree structure that supports the atrium of the University of Guelph Science Building in Guelph, ON, Canada, designed by Young + Wright Architects with Walters Inc., required full surface remediation to blend between the cast nodes at the intersections and the mechanical pipe that forms the branches. No welded connections are visible.

4.4 Weld show-through must be minimized. The dimpling on the back face of the welded element caused by the heat of the welding process can be minimized by hand-grinding the backside of the weld. The degree of weld show-through is a function of weld size and material.

The custom-curved hollow sections for the Peace Bridge in Calgary, AB, Canada, designed by Santiago Calatrava, are reinforced on the interior with plate sections. The heat impact of this welding is just barely perceptible on the connection, indicating minimization of the effect.

AESS C – CUSTOM ELEMENTS

AESS C – Custom Elements was created to allow for a completely custom selection of any of the characteristics or attributes that were used to define the other categories. It allows complete flexibility in the design of the steel, but therefore requires a high level of communication among the architect, engineer and fabricator. The cost premium for this type of AESS could range from 20% to 250% over regular steel. A wide range of variation may seem odd for custom elements, but the lower bound of this category also includes specialty reused steel for sustainable purposes, and steel that might be purposefully less refined in its characteristics.

USING THE SYSTEM

It is most efficient to work through the decision-making process of an AESS project using the Categories and Characteristics System from the beginning. In a few instances fabricators have requested that the architect and engineer incorporate the AESS Category System at the bid phase of a project, to make the bid process smoother and to put all fabricators bidding the project on equal terms.

It is strongly recommended that the team sit down with the matrix (next page) and specification documents in hand, and manually go through the list of characteristics. As is illustrated by the range of projects shown here, the complexity, size, level of finish and types of members can greatly vary in Custom AESS projects – leading to a wide variation in the cost premium to be expected for this type of project.

Specific areas of concern for Custom AESS projects might include:

- oversized members
- extraordinary geometries
- curved members
- incorporation of tensile systems
- long-span structures
- accessibility issues
- unusual finish requirements
- high levels of grinding and filling for connections
- transportation problems associated with member size
- difficult handling or extra care needed to protect pre-painted components

Many variations of these and other connection and design-based issues will be addressed in the following chapters.

AESS 1

AESS 2

AESS 3

AESS 4

AESS Category Matrix

	Characteristics
1.1	Surface preparation to SSPC6 or AS1627 Sa2/Class 2
1.2	Sharp edges ground smooth
1.3	Continuous weld appearance
1.4	Bolt head orientation specified
1.5	Weld spatters removed
2.1	Visual Samples
2.2	One-half standard fabrication tolerances
2.3	Fabrication marks not apparent
2.4	Welds uniform and smooth
3.1	Mill marks removed
3.2	Butt and plug welds ground smooth and filled
3.3	HSS/RHS/CHS weld seam oriented for reduced visibility
3.4	Cross-sectional abutting surface aligned
3.5	Joint gap tolerances minimized
3.6	All-welded connections
4.1	HSS/RHS/CHS weld seam not apparent
4.2	Welds contoured and blended
4.3	Surfaces filled and sanded
4.4	Weld show-through minimized
C1	
C2	
C3	
C4	
C5	
Sample use	
Estimated cost premium	

AESS C	AESS 4	AESS 3	AESS 2	AESS 1	SSS
Custom Elements	Showcase Elements	Feature Elements	Feature Elements	Basic Elements	Standard Structural Steel
		Viewed at a Distance ≤6m/20ft	Viewed at a Distance >6m/20ft		
	×	×	×	×	
	×	×	×	×	
	×	×	×	×	
	×	×	×	×	
	×	×	×	×	
	OPTIONAL	OPTIONAL	OPTIONAL		
	×	×	×		
	×	×	×		
	×	×	×		
	×	×			
	×	×			
	×	×			
	×	×			
	OPTIONAL	OPTIONAL			
	×				
	×				
	×				
	×				
Elements with special requirements	Showcase or dominant elements	Airports, shopping centers, hospitals, lobbies	Retail and architectural elements viewed at a distance	Roof trusses for arenas, retail warehouses, canopies	
Low to high (20–250%)	High (100–250%)	Moderate (60–150%)	Low to moderate (40–100%)	Low (20–60%)	None 0%

ERECTION CONSIDERATIONS

CHAPTER 4

TRANSFORMING AN ARCHITECTURAL IDEA INTO PREFABRICATED AESS ELEMENTS

A diagrid section is lifted onto the base structure of the World Financial Center entry pavilion in New York City, NY, USA, designed by Pelli Clarke Pelli Architects, with fabrication by Walters Inc. and erection by Metropolitan Walters. The components had been test-erected in the shop to ensure a good fit prior to transport. The complexity of this structure and the desire for a seamless AESS 4 finish required special precision in the connections.

There is significant work involved in transforming the architectural idea of a steel-framed building into a series of prefabricated elements that can be readily erected on the building site. Even into the 21st century, and in spite of advances in technology, the design, fabrication and erection of steel buildings remains a hand-crafted process. This is even more true when using AESS due to the additional aesthetic requirements. There is human interaction, workmanship and decision-making during every step. The individual AESS project is quite unique so that the design, fabrication and construction must to a large extent be customized to suit the project. This might seem to run counter to the idea behind the early development of iron and steel as being suited to mass fabrication and assemblage construction, while the industry still relies on these precepts as the basis for achievements in economy and speed of erection. Aspects of pure craft and pride in workmanship remain core to steel design.

Where the architect's proposal begins to stretch the limits of the use of successful precedents in detailing, fabrication and erection, fabricators are often brought into the discussion ahead of the finalization and tendering/bid phase of the project, to inform the detailing. At a later stage, the ironworkers who erect a project are critical to its proper completion. In all projects, but particularly when AESS is used, the problem-solving skills of the lead ironworker can make or break the pace, speed of erection and timely completion of the work. The ironworkers will have a sense or feel of the fit of pieces. They will be responsible to ensure the proper alignment during erection and the quality of finish of any site welding and finishing operations. This is also a dangerous job. Even if tied off and wearing proper fall protection, much of the work is done at a great height, during all sorts of weather and around moving elements that may weigh thousands of tons. Aesthetically driven decisions regarding the predominance of site bolting or welding feed into aspects of safety and timing on the site.

SHOP VERSUS SITE FABRICATION

As will be discussed further in *Chapter 7: Connections*, the design of the steel system will require critical decisions on welded or bolted appearance. For efficiency and cost reduction, it is always preferable to maximize welding and finishing operations that can be done in the shop, which is a more controlled environment and where the elements can be lifted and turned to make them more easily accessed for working. Welding quality tends to be higher in the shop, although this does not preclude quality site welding, which is more difficult and time-consuming and will depend largely on the abilities of the site crew. This means that the design of AESS benefits from a higher than normal level of shop fabrication as it will be easier there to achieve high-quality finish results. This infers that the sizes of the elements that are being fabricated and potentially painted in the shop should be maximized in order to minimize, in particular, site-welded connections. This creates a difference in role and nature between the connections fabricated in the shop and those that are completed on site. Site connections need to be designed to acknowledge erection constraints. This may infer the need to complete connections more swiftly – which leads to a preference for bolting over welding. Highly articulated welded connections need to be done in the shop.

The segments of the AESS diagrid columns for the World Financial Center entry pavilion were test-assembled in the shop to ensure a trouble-free erection. This procedure was essential due to the complex geometry of the project and the need to subdivide the structure into elements small enough to be transported by truck. A driving design factor in this project was the desire for all-welded connections so that the curved vertical tubes would be seen as if they were single elements. As they are close to five storeys high, this was physically impossible due to transportation constraints. The requirement of AESS 4 added greatly to the extremely precise fit required for the welded connections between the tubes, which were carried out on site.

The welding for the World Financial Center entry pavilion steel was more easily carried out in the shop using lifts. Yet the complex geometries of the steel made access difficult for some parts. Note that the tube-to-tube splices are located at a distance above the horizontal plate to provide access to the detail for welding.

THE IMPACT OF TRANSPORT ISSUES ON DESIGN

The maximization of the work carried out in the shop may mean that members can become increasingly large and difficult to transport. It will be essential that the fabricator map the clearances from the shop to the site to ensure that the pieces will fit for easy transport, including turning radii for narrow streets and bridge clearances. It is obviously less expensive to avoid requiring an escort or street closures. Load capacities of roads and bridges will also come into consideration.

The elements for the diagrid columns for the World Financial Center entry pavilion had to be shipped over 700km/435mi. Entry to the island of Manhattan is only possible by bridge. Special permits needed to be obtained for the transport at a very specific time during the night, so as not to obstruct traffic flow.

To prevent damage, members may have to be shipped separately, rather than maximizing the allowable tonnage per trailer as would be normal for concealed steel. More delicate members may require the use of temporary steel bracing to prevent distortion from road movement, offloading and subsequent lifting.

Individual pieces of steel cannot be any larger than can be transported to the site. This limit in size will determine the placement of the connections between pieces: the more connections can be shop-fabricated, the more economical the pieces, and the quicker the erection. If the project aesthetic is for all-welded connections, then further thought has to be given to how the site welding will be accomplished during the erection process. Preference is given to making the pieces fit on a standard trailer, as most large fabricators will own a fleet of these. Custom trailers or sets of wheels can be fabricated especially for the project.

The 27m/89ft-long legs for the Ontario College of Art and Design in Toronto, ON, Canada, designed by Will Alsop Architect, fabrication and erection by Walters Inc., used a split trailer for transportation as the legs, which were shipped two at a time, exceeded the length of a standard trailer. The site was situated in a highly constricted urban area with narrow streets.

The sections of the World Financial Center entry pavilion columns needed to be broken into numerous smaller elements for shipping. This meant a much lighter load for transport, and one that had to be fitted with temporary bracing to keep it stable during shipping. This also allowed for the pieces to be offloaded in a vertical position, which saved them from damage and distortion and put them in the correct orientation for subsequent lifting. The height of the pieces was determined in part by overhead clearances.

The permitted member size will also be impacted by the amount of staging area for the project. Even if the member size is limited due to transportation, some controlled sub-assembly can take place in the staging area for the project prior to lifting. The site crane can be used to lift and rotate the members in the staging area in order to allow ground access for the ironworkers to complete the connections.

SITE CONSTRAINTS

The available staging and offloading area will impact the erection process. Adequate room for the different capacities and functionalities of cranes is required. The larger cranes are sized according to their lifting capacity and reach. Temporary cranes are often brought to the site to lift larger special steel elements. These need a large area to accommodate the outreach of the stabilizing feet. The ground beneath them must be able to bear the weight of the crane at its full lift capacity. Smaller cranes or lifts will be required to allow access to the pieces for crane release as well as connection operations. The AESS construction is often not the only simultaneous activity on the site and interference with other operations needs to be coordinated.

Constricted sites are common in dense urban areas. Lane closures may be required on fronting streets to provide for staging and erection, particularly when building to the lot line. Street closures to accommodate large lifts are liable to push erection times to night hours in order to minimize impact on traffic. These sorts of timing limitations call for connection types that can be completed within a short time frame. This concerns both the permanent bolted connections and temporary bolting that puts the steel into a structurally stable, self-supporting state while waiting for the completion of the final welding operations.

This urban construction site with severe access restrictions required very careful threading of a smaller crane to be lifted in to assist with the erection process.

For oversized or geometrically complex members it is not uncommon for sub-assembly to occur on site in the staging area. The size of the staging area will figure into design decisions that affect the types of connections employed in aggregating very large members. Quality welding in the staging area may require the use of jigs. If an all-welded appearance is desired, the design may need to make use of inventive "hidden" bolted connections to ease erection. This detailing will be described in *Chapter 8: Specialized Connections*.

Site constraints are also critical in renovations and additions, in particular where the operation of the project must continue throughout the construction process, which can also mean that erection hours may be limited.

The extensive renovation of Union Station in Toronto, ON, Canada, designed by Zeidler Partnership with steel fabrication and erection by Walters Inc., had as its requirement that the station be kept fully operational for the duration of the project. Most of the work had to happen at night. The single primary crane was positioned such that it could fully reach all lifts, as the constricted urban site could not accommodate other locations for cranes. It is also more cost-effective to limit the number of cranes. AESS located close to the ground was protected with plywood to prevent damage from equipment or patrons.

The same type of software that is currently used by fabricators to assist in the detailing can also be employed for setting out the staging of projects.

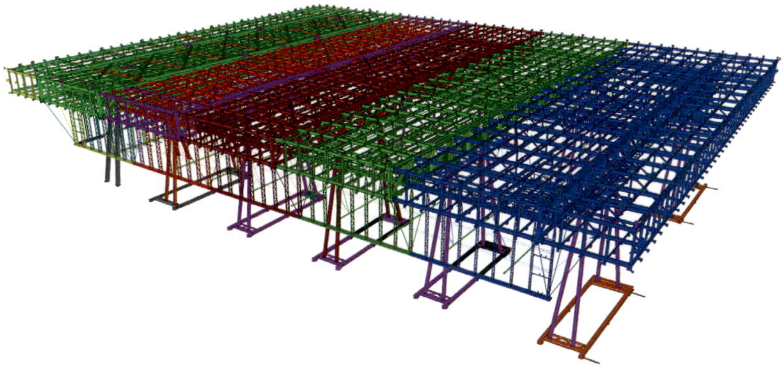

This digital drawing of Union Station by Walters Inc. was prepared with Tekla Structures. It demonstrates the size of the facility and the sequence of construction.

CARE IN HANDLING

AESS requires more care in handling to avoid damage to the members. Although this largely references the surface characteristics, it can also concern distortion. Oddly shaped or eccentric members can easily be distorted or bent if improperly handled. Many of the members that come to the site may also be pre-finished with paint, galvanizing or intumescent coatings, so that padded slings will be required to avoid marking the finish coat. Lifting can be done using nylon straps for lighter member or padded chains for heavier elements. The more precisely fabricated the pieces are, the less force will be required to fit them during erection. This is one of the intentions of the AESS characteristics, as discussed in *Chapter 2*, that ask for tighter tolerances.

Many AESS components will require temporary supports during the erection process. If shoring "rests" are used, the surface of the rests will require padding with pieces of carpeting or rubberized materials to avoid steel-to-steel contact.

Heavily padded chains are used to manoeuver the steel for the World Financial Center entry pavilion (top). This was to avoid damage to the future AESS 4 finish requirements.

The installation of the tubular branches for the central tree-like atrium support system at the University of Guelph Science Building in Guelph, ON, Canada, designed by Young + Wright Architects with Walters Inc., (right) used fabric slings to minimize damage to the surfaces during erection as well as padding in the seat of the temporary support system. This structure was to receive a high-gloss finish, so no damage could be tolerated to the surface of the steel.

As the legs for the Ontario College of Art and Design were erected with their final intumescent coating in place, more care than normal was required during the erection process. The blue temporary support frame was custom-fabricated and the curved and padded support sized to fit the tube.

Even the extremely sturdy-looking diagrid nodes for the Bow Encana Tower in Calgary, AB, Canada, designed by Foster + Partners with fabrication by Walters Inc., employ stabilizing steel to ensure that the precise geometry and alignment does not shift during transport. As the face of the node will be finished to AESS 4 standards, careful site remediation was required after the braces were cut away.

SEQUENCING, LIFTING, ACCESS AND SAFETY

Just-in-time delivery is needed to ensure proper sequencing. The erector will arrange lift sequences to minimize the amount of steel that is on the site at any time. Construction sequencing for AESS elements places further limitations on detailing and increases the challenge of erection.

One of the costs that can increase dramatically with AESS is associated with the number and variety of cranes, person lifts and temporary platforms required to both erect and finish the steel. Where bolted connections can be completed fairly quickly and safely from a lift, intense welding usually requires the construction of temporary platforms. If work must continue through inclement or cold weather, work platforms must provide additional protection. The standards for this vary globally, but it is the contention of this author that designers must be aware of the ramifications of design detailing and erection issues as they impact the health and safety of the workers.

Restricted site access, a densely built-up site and the angled steel supports on QRC West in Toronto, ON, designed by & Co Architects, make this a logistical and sequencing challenge. A large crane to the back of the site lifts the steel leg. A smaller lift is required to access the connections to complete the temporary bolting. Fabrication and erection by Walters Inc.

Often streets must be closed off from traffic in order to provide better access to the site for the erection of temporary cranes that complete significant lifts. As this delta frame for QRC West is located directly adjacent to a fairly narrow street, the work had to be scheduled after 8pm to limit the impact on traffic. Night work is more difficult and costly for obvious reasons. Casting designed by CastConnex.

The construction of the International Terminal at Calgary International Airport in Calgary, AB, Canada, designed by DIALOG with fabrication and erection by Supermétal (top), illustrates the number and range of lifts required to complete work on the connections. The load capacity of the floor must be considered when selecting equipment.

Erection at the site of the World Financial Center entry pavilion (left) was complicated by site conditions. The steel was being erected over the access to the underground PATH public transport system and the lifts had to be light enough to operate on the newly installed concrete floor system. The larger crane had to be located on a higher load capacity portion of the site and sized so that it could reach both diagrid columns. The roadway adjacent to the site was under reconstruction and access in general was extremely restricted due to the extent of construction of the new World Trade Tower opposite.

The selected AESS category feeds directly into project costs as it will impact both crane and lift rentals as well as the labor involved to complete connections. Bolted connections are able to reduce these time-based costs while welded connections will increase costs due to the construction of additional temporary platforms. The concept of distance to view, which is a primary determining factor in category choice, was defined to assist designers in natural efficiencies and economies.

Decisions need to be made that take into account:

- maximum length of member that can be transported
- span length of members
- location and requirements for splices and connections
- crane capacity and location
- staging area
- access to the site for erection

The trusses that span across the departures hall of the International Terminal at Calgary Airport were too long to be transported in one piece. In addition to the use of hidden bolted connections, another splice was necessary that required welding of the tubular members. The nature of the site allowed for this connection to be fully completed prior to lifting, saving time and cost.

An ironworker is completing a high-quality welded splice within the long truss member in advance of the lift. Remediating to AESS 3 and higher categories is very painstaking and time-consuming work. A hidden splice connection would have been an acceptable alternate solution.

Large AESS projects in China and the Middle East are making predominant use of concrete-filled steel tubes, which requires extensive welding of member-to-member connections on site. As a result there is more welding work done at height than might be seen elsewhere. Different materials, such as bamboo, may be used instead of steel pipe to create scaffolding and platforms. Bamboo scaffolding is used extensively in China as the material is plentiful, renewable and extremely strong.

Capital Gate in Abu Dhabi, UAE, designed by RMJM Architects, used a welded hollow structural frame to create its architecturally exposed diagrid structure. Platforms were erected at each of the nodes to create the final welded connections, which needed to be remediated to the point of becoming invisible. The structure was given an intumescent fire-protective coating that concealed some of the connection work.

ERECTING THE STEEL

The process of erecting AESS varies with the complexity of the project. If the steel members have been accurately constructed with no less than half standard tolerances, fitting issues should be minimized but may not have been eliminated. The maximization of shop fabrication and transport size is intended to result in a minimization of site connections. It is not unusual to see the erection of extremely large members, which impacts the crane capacity requirements.

Bamboo scaffolding is used on the Canton Tower in Guangzhou, China, designed by IBA Architecture and Arup. The exposed concrete-filled diagrid tube framework extends to a height of 600m/1,969ft.

As inspection of the exposed steel of the Canton Tower (bottom left) is carried out high above the city of Guangzhou, China, ironworkers wear fall protection attached to a system of guide ropes attached to the structure.

Bamboo scaffolding is erected (bottom) high above the city of Guangzhou to provide access to the spiraling exposed diagrid of the Canton Tower. As much of the tower consists of open structure, with minimal contact with floors, significant work had to be carried out via climbing through this framework. Fall protection gear is clipped to the scaffold frame.

Lifting points on the members are calculated to put them at the correct angle of approach when lifted for bolting or welding. For normal orthogonal steel erection, gravity assists in pulling the pieces into their final position. Standard structural steel elements tend to be more regular, with vertical columns and horizontal, relatively uniform beams. In such cases the lifting points are predictable and make assembly on site routine and quick. Columns have an attaching point at the top to be hooked onto the crane. Symmetrical beams can be lifted from one central point for short members or via a sling for longer or slightly asymmetrical members.

With odd geometries and asymmetry of members, the lifting points will need to be more carefully pre-calculated. With diagonal or unbalanced members, gravity will not be of assistance and lifting points may require more calculation than normal. Angled or unevenly loaded members will have custom-designed lifting chains of uneven lengths that are worked out by the fabricator/erector in advance of the lifting to minimize delays during the erection process.

Extreme AESS projects will include those with little consistency in member design and geometries. There may be erection delays in projects where each member is unique. It is not unreasonable for some members to require more than one attempt due to alignment issues.

The base element of the World Financial Center entry pavilion diagrid column is lowered onto its foundation. This is the largest of the elements, based on the limits of truck transport to the site. The eccentricity of the shape of the column is minimal towards its base, so it does not greatly affect the lifting points and leveling of the base plate in its alignment with the receiving bolts. As all of the other connections have been tested in the shop for fit, this is the only potential element of surprise for this erection sequence.

The tubular legs of QRC West are angled, so the lifting points, chain lengths and crane attachment were designed to take this into account. The concrete bases have additionally been angled to provide a more direct load path. A temporary support system was erected to hold the angled leg in place until stabilized by the addition of other permanent members. Ropes are attached to the pieces to assist the ironworkers on the ground to guide the steel.

The angle of approach of the element must be precise in order for it to land properly and align with the foundation bolts. It is nearly impossible to force extremely heavy members to align if this is not properly set out. Given the round plate, it was also necessary to carefully mark the piece to ensure the correct rotation. The round tube visible at the base of the tapered leg is to facilitate filling the structure with concrete.

COMBINING AESS WITH OTHER SYSTEMS

When Architecturally Exposed Structural Steel is used in conjunction with timber and glass systems, tolerances become even tighter and the erection process more complex. These topics were discussed in detail in *Understanding Steel Design: An Architectural Design Manual* in Chapter 12, "Steel and Glazing Systems", and Chapter 13, "Advanced Framing Systems: Steel and Timber".

Ironworkers at the Art Gallery of Ontario in Toronto, ON, Canada, designed by Frank Gehry, are erecting a glue-laminated timber beam that is fitted with galvanized steel connections. The wood products require an even higher level of care in handling than their AESS connection elements. Any force put on the member could cause it to split. Steel fabrication and erection by Benson Steel.

COATINGS & PROTECTION

CHAPTER

5

The structure of the Beijing National Stadium ("Bird's Nest") in Beijing, China, designed by Herzog & de Meuron/Ai Weiwei/Arup, 2008, was to a large part finished with metalizing after welding. Complex joints and surfaces that could not accommodate metallization were finished with a series of coatings of field-applied materials. The surface was first roughened, followed by two coats of "Zinga", a spray-applied intermediate layer of Epoxy Micaceous Iron Oxide (MIO), followed by also spray-applied metallic-grey fluorocarbon finish.

Coatings on AESS are critically important as they impact durability. SO_2 and NO_X, by-products of high levels of traffic and certain industries and known in some regions as acid rain, can react with certain finishes causing accelerated degradation, as in the case of galvanization.

Surface preparation of the steel is the primary means to ensure proper adhesion and quality of varied finishes on all categories of AESS. This includes designing to shed water, prohibit corrosion and provide adequate layers of coatings to ensure that they last and do not wear thin in high- traffic areas.

The finish and the AESS itself must be designed to be maintained and cleaned. Choices in the color will impact cleaning frequency. The use or function of the building should inform the choice of color. Train stations, for instance, are more prone to degradation from airborne particulates. This holds for any exterior application of steel, including canopies, bridges and car park covers.

The exposed steel roof of the Southern Cross Rail Station in Melbourne, Australia, designed by Grimshaw Architects, is finished in a medium grey, which is excellent at masking evidence of grime.

The exterior structure of Wembley Stadium in London, England, designed by Foster + Partners, shows evidence of dirt build-up due to rainwater runoff on the underside of the steel frame.

Although finishes are applied on the completed steel, it is imperative that they are selected and known at the outset of the project. All aspects of exposure and detailing will need to be taken into account when defining the nature of the finish. It is crucial that a fire code review be conducted to determine if the steel can indeed be exposed, and if fire protection is required.

Finishes vary from thin to thick and from matte to glossy. The choice of finish, as well as its color, will impact the visibility and durability of the detailing. If a high-gloss finish is desired, it will reveal every minute imperfection in the steel. Flat finishes are more accommodating. Light-colored paints will quickly reveal corrosion and dirt. Thin finishes will reveal surface imperfections. Thicker coatings such as intumescent fire protection can cover or conceal imperfections but fine details as well.

SURFACE PREPARATION

Surface preparation will be done in accordance with the chosen AESS category. At the minimum, as specified in AESS 1, the material needs to undergo commercial blast cleaning to remove any grease, paint and other materials that will prevent the proper adherence of finishes. If the surface is not adequately cleaned prior to the application of the coating, the coating system is likely to either fail or the surface deficiencies will show through. It is important to note that the members may need to be cleaned more than once and by different methods during the fabrication and erection process. If elements undergo commercial blast cleaning and are primed immediately, they will not have the opportunity to oxidize. Portions of elements that are to be welded cannot be primed, therefore these areas will require sanding and cleaning immediately prior to welding to bring the surface to a suitable condition.

Where different AESS categories are used in the project, there may also be different surface preparations and different finishes required. This will be especially true where different varieties of steel are joined in one detail. Standard hot-rolled carbon steel, resin-cast steel and mechanical pipe all have inherently different surface textures.

The diagrid baskets at the World Financial Center entry pavilion that arrive at the site are primed, but the primer had been held back from the connection points to be welded. These abutting surfaces have to be ground and prepared to ensure that they are free of oxidation and also milled flat to ensure the best fit and welded strength.

The surface of this large resin-cast node (left) is akin to an orange peel. To join this to components that have an inherently different, smoother finish requires thoughtful detailing if excessive remediation is not in the budget.

Connections can be detailed so that the surfaces are separated, thereby requiring far less remediation. This was done for the large cast nodes at King's Cross Station in London, England, designed by John McAslan and Partners with Arup.

Finishes for exterior steel structures will need to pay special attention to the prevention of corrosion. Paint will not make up for design deficiencies. Even the most sophisticated epoxy and vinyl paint coatings cannot compensate for details that create opportunities for corrosion to occur. The basic selection of member type and connection detailing for exterior structures should ensure that there are no places where water can collect or puddle. With some care and attention, orientation problems can be overcome. Beams and channels should be oriented with the webs vertical so that water cannot collect and stand for any period of time. Exposed steel on which moisture can collect should be detailed with a slope to ensure drainage. Drain holes should be added if the section cannot be oriented or sloped to drain.

When using hollow sections or composite elements that create voids on exterior applications, it is necessary to prevent corrosion also on the interior surfaces. Often seal welds are specified to prevent the entrance of moisture or oxygen-laden air into the cavity. Seal welds may be specified to prevent unsightly rust bleeding on AESS that is to be painted. The seal weld prevents moisture movement, it does not add strength to the joint.

PAINT SYSTEMS

Steel exposed to view is generally painted for appearance unless a higher level of fire protection is required. A one-coat paint system can be sufficient for standard warehouse structures that will not be top-coated (standard structural steel and AESS 1). Since the building environment is controlled, no corrosion occurs once the building is enclosed.

Steel exposed to view that will be top-coated for appearance (AESS 2 and above) requires a prime coat for adhesion. A fast dry primer is sufficient to provide the necessary base. To ensure that this system will perform for long periods a greater degree of cleanliness is required by the specification. AESS standards require surface preparation to a minimum of commercial blast cleaning. Consultants must ensure that the finish coats are compatible with the primer. Alkyd primers are acceptable but epoxy primers are not.

Structural steel that is exposed to view and weather on the exterior of buildings requires more thorough cleaning and finishing to ensure long-term performance. Higher degrees of cleanliness along with better-quality multi-coat paints should be considered under these circumstances. Epoxy systems over compatible primers are usually most suitable. Urethanes should be used when wear is a consideration.

Tender/bid documents should include the following information to ensure good-quality coating systems:

- identification of the members to be painted
- a specification for the degree of cleanliness required to ensure performance such as SSPC Surface Preparation Standards
- compatible primer, intermediate and finish paints and if applicable
- the manufacturer's product identification
- the average dry film thickness per coat

PRIMERS

Not all finish systems require a primer. The selection of the primer will be a function of the choice of the finish coating. Not all finish coating systems take the same base primer, so revisions in the final finish type may require remedial correction of primers to ensure compatibility. Care in application of the primer is important as any drips and runs will translate through both paint and intumescent coating finishes.

Some AESS projects that will be exposed to the weather for long periods during construction will choose to use a primer to prevent corrosion and subsequent re-cleaning of the steel prior to the application of the final finishes.

The primer is held back from the edges of the welded connection as it is not possible to weld over a finish. The corrosion that has occurred will be ground away immediately prior to welding.

SHOP VERSUS SITE-APPLIED COATINGS

The painting and finishing of an AESS structure can take place in the fabrication shop or on the site. Many fabricators can offer shop-painting that can ensure a more consistent, higher-quality finish. Shop-applied paint finishes likely need to be touched up after erection, but this is usually less problematic than the complete painting of the structure on site. Pre-painted structures also require extra care and protection during transportation, handling and erection. Pre-painted structures will be more in need of just-in-time delivery to the site to prevent site-generated damage. For the same reason, pre-painted structures may also require better staging areas on site.

The careful preparation of the steel, including the basic removal of sharp edges, allows for a more even application of the paint and better coverage on the corners. The spray application of the product on sharp corners is difficult and if these are not ground or slightly rounded off, premature wear on the edges of the structure may occur, particularly in high-traffic locations, and corrosion on exterior applications. The same concerns hold for the application of intumescent coatings (see page 89). In addition, some of these coatings have a preference for either shop or site application, due to their materiality, VOC content and application process.

In multi-coat systems it is essential to provide adequate drying time between layers. Failure to do so will lead to accelerated degradation of the coating system due to the entrapment of uncured ingredients.

CORROSION PROTECTION

When steel is exposed to weather, coatings must be applied that prevent its natural oxidation. The method chosen must reflect the specific climate and weathering issues, the type of structure, the category of AESS (including its budget) and the desired nature of the detailing. Rain, snow, atmospheric pollution and coastal and marine environments may all cause oxidation acceleration. Additionally, the use of de-icing agents such as salt can be quite disastrous, even if reasonable care has been taken in the detailing and selection of a coating. The most common choices for coatings are galvanization and metallization. An alternative would be to use stainless steel or weathering steel in lieu of coating regular carbon steel.

Galvanization

Galvanized finishes are increasingly seen in AESS applications. It is therefore important to remember that in the view of the steel industry, galvanizing was not intended to be a finish but a preventative measure against corrosion. The speckled grey finish will inevitably vary from batch to batch, even within the same manufacturer and also as a function of the application technique and the style, size and shape of the member to which it is being applied. The variation will dull down over time as the finish ages but will not disappear altogether.

Galvanizing has been used extensively as both corrosion protection and final finish for the Calgary Water Building in Calgary, AB, Canada, designed by Manasc Isaac Architects. The coloration is different on the two segments of the round HSS column. A detail that separates the two sections makes this difference less apparent. The bow-shaped wide-flange truss has been installed in components, the parts sized to fit in the galvanizing bath. Even the decking uses a galvanized finish to make for a very consistent appearance.

Achieving a good-quality coating requires that the surface be free of grease, dirt and scale on the iron or steel before galvanizing. When the clean steel component is dipped into the molten zinc (about 450°C/840°F), a series of zinc-iron alloy layers is formed by a metallurgical reaction between the iron and zinc. When the reaction between iron and zinc is complete, there is no demarcation between steel and zinc but a gradual transition through the series of alloy layers that provide the metallurgical bond. This helps to make the galvanized finish highly durable as it cannot easily be chipped away. As the thickness of the coating is determined by the thickness of the steel, the galvanized coating can be made thicker by roughening the steel, thereby creating more surface area for the metallurgical reaction to take place.

Galvanized coatings protect steel in three ways:

- The zinc weathers at a very slow rate, giving a long and predictable life.
- The coating corrodes preferentially to provide sacrificial protection to small areas of steel exposed through drilling, cutting or accidental damage.
- If the damaged area is larger, sacrificial protection prevents sideways creep that can undermine coatings.

No post-treatment of galvanized articles is necessary. Paint or a powder coating may be applied for enhanced aesthetics or for additional protection where the environment is extremely aggressive.

The resistance of galvanizing to atmospheric corrosion depends on a protective film that forms on the surface of the zinc. When the steel is lifted from the galvanizing bath, the zinc has a clean, bright, shiny surface. With time this changes to a dull grey patina as the surface reacts with oxygen, water and carbon dioxide in the atmosphere. The result is a tough, stable, protective layer tightly bonded to the zinc. Contaminants in the atmosphere will affect this protective film. The presence of SO_2 greatly accelerates the atmospheric corrosion of zinc.

Where AESS is being installed in an exterior environment, it is critically important that all surfaces be coated. Complex shapes and most hollow items can be galvanized inside and outside in one operation. For complete protection, molten zinc must be able to flow freely to all surfaces of a fabrication. With hollow sections or where there are internal compartments, the galvanizing of the internal surfaces eliminates any danger of hidden corrosion during service. Good member design requires the means for the access and drainage of molten zinc and the means for escape of gases from internal compartments (venting).

It is important to bear in mind that the steelwork is immersed into, and withdrawn from, a bath of molten zinc at about 450°C/840°F. This temperature can cause distortion in thinner steels. If the use of the galvanized coating is known early on during the design process, it may be decided to increase the thickness of the steel to prevent distortion. From a design perspective, it is important to understand the physical limitations of the galvanizer's facility. It is not unusual to dip pieces that are 20m/66ft in length, but this limit must be verified as it impacts member size. This limit on the member size may result in the need for additional connections.

Any features that support the access and drainage of molten zinc will improve the quality of the coating and reduce costs. With certain fabrications, holes that are applied for other purposes may fulfil the requirements for venting and draining; in other cases it may be necessary to provide extra holes for this purpose. Seal welds may be specified to prohibit pickling acids and/or liquid zinc from entering into a specific region during the galvanization process.

This pedestrian bridge in Olympia, Greece, uses durable galvanized steel for its structural and railing components. Yet some of the bolts appear to be plain carbon steel and are oxidizing. This is easily avoided by specification of galvanized fasteners.

Seal welds are liable to alter load paths and are prohibited in some structural situations. It might be better to provide a vent space and also galvanize the interior of hollow sections. This will increase costs directly due to the increased surface area to be galvanized but will provide a more durable exterior coating. Proper communication is important when deciding on the method of prevention of moisture entry on sealed joints.

The corrosion finish on the element must be matched by the type of connector. It is extremely important to use connectors whose finishes are durable (even with wrench tightening) and that will not rust. Metals must be chosen that are chemically compatible.

Metallization

Metalizing is a substitute for painting structural steel. It protects steel for significantly longer than paint alone. Steel of every shape and size may be metalized either in the shop before construction or on site. Metalizing creates a very versatile and effective coating for steel structures that are continuously exposed to weathering.

The metalizing process begins with proper surface preparation. Next, aluminum wire or zinc wire is continuously melted in an electric arc spray or gas flame spray gun. Clean, compressed air strips droplets of molten metal from the wire, depositing these particles onto the steel and thereby forming the protective coating. This sprayed metal coating is both a barrier coating and a galvanic coating in one. A single metalized coating can protect steel for 30 years or longer, depending upon the application, coating thickness and sealing.

Metalizing is thought of as a cold process in that the aluminum or zinc is deposited onto the steel by spraying, rather than by dipping the steel into a bath of molten zinc. The steel remains relatively cool at about 120°-150°C/250°F-300°F. This means that there is virtually no risk of heat distortion by metalizing. There are no VOCs in the metalized coating. There is also no cure time or temperature limit to metalizing, so that this process may be applied throughout the year, virtually regardless of temperature.

Three different types of wire are used to create three specific coatings. Sprayed aluminum is preferred for use in industrial environments, particularly where there are high concentrations of sulfur dioxide and other pollutants. Zinc provides greater galvanic protection than aluminum. Its greater galvanic power protects gaps in the coating better than pure aluminum. It is marginally easier to spray pure zinc than pure aluminum by some flame or arc spray systems. Zinc with 15% aluminum wire combines the benefits of pure zinc with the benefits of pure aluminum in the metalized coating. This mix is often used as a substitute for pure zinc, because it is somewhat more chloride and sulfur-dioxide resistant than pure zinc, while retaining the greater electro-chemical activity of pure zinc.

WESTGATE PEDESTRIAN AND CYCLE BRIDGE | AUCKLAND, NEW ZEALAND

Jasmax architects, Aurecon Consulting Engineers and HEB Construction, 2013

Auckland has a fairly corrosive climate due to its adjacency to water. The Westgate Pedestrian and Cycle Bridge made extensive use of metallization as the basis for the corrosion protection strategy. The complex geometry of the trusses and skewed piers required the use of 3D drafting and 3D structural analysis models for the analysis and seismic design of the structure.

The 100m/328ft ramp was curved into a helix that achieved a low ramp slope ratio of 1:13. The tubular truss was positioned above the pedestrian deck, which addressed clearance issues over the highway and lowered the deck height by 2m/6ft. This in turn reduced the length of the ramp.

The yellow truss structure was fabricated by D&H Steel from circular hollow sections. The steel was given a metalized coating in their shop, followed by a primer. The top coat was also shop-applied to ensure a high quality and consistency of finish. This required greater care in handling and erection. The connections between the tubular truss members were welded, without any remediation of the welds prior to the application of the coating system.

The fabrication of the upper truss portion of the bridge in the D&H Steel shop. All of the welded connections are completed in the shop. The welds between the sections of the top chord are visible, while the remediation to these splices is evident as there was the desire to make the transition seamless. Only the lower portion of the bridge is given the metalizing coating, as this portion has more exposure and potential for corrosion than the upper section. Care is taken to ensure that the welds are sealed to prevent moisture from entering the tubular steel.

A primer is applied over the metalizing to provide a base for the final paint coating. The selection of the primer will depend on the requirements of the final paint finish or coating. The welds have been neatly done and left exposed at the panel points of the truss.

A close view of the surface of a welded connection after the application of metalizing.

The pedestrian bridge (here under construction) is given final paint coatings to provide the architectural finish.

Weathering Steel

Weathering steel has the unique characteristic, under proper conditions, to corrode to form a dense and tightly adhering oxide barrier that seals out the atmosphere and retards further corrosion. This is in contrast to other types of steel, which form a coarse, porous and flaky oxide that allows the atmosphere to continue penetrating the steel. The addition of chromium, copper and nickel alloying elements give the weathering steel its enhanced resistance to corrosion. The oxidized layer on weathering steel in many climates does not consume a significant amount of steel in its formation. However, climate is important – the oxide layer will form only under the condition that there are wet/dry cycles.

After about two years, which is about the time it takes to develop the oxide skin, the color of weathering steel has turned much darker and reddish brown. As we are dealing with "living steel", the color will not be consistent from project to project or even within an individual project. In wetter climates, the color of weathering steel will generally have an overall redder cast compared to those exposed in drier climates. However, one can be most certain that the final color will have a rich dark earthy tone. It is low-maintenance, durable and beautiful, provided that the design has been careful about the details.

Special attention must be paid to the drainage of storm water (or condensate) to prevent staining of surrounding structures, sidewalks, and other adjacent surfaces. Runoff of water from upper portions of a structure tends to produce long-lasting streaks or other patterns of redder oxide on lower portions. Marine environments are not suggested for this material. Exposure to high concentrations of chloride ions, originating from seawater spray, salt fogs or coastal airborne salts, will cause accelerated rusting. It is suggested that weathering steel not be used closer than 2km/1.24mi of the shore, although it is noted as being variable as a function of the analysis of the direction of wind-borne moisture. Excessive use of de-icing salt can also lead to accelerated corrosion, as can atmospheric pollution, given that weathering steel reacts badly to high concentrations of SO_2.

Weathering steel is not readily available in other than plate goods. The absence of hot-rolled sections or tubular material means that all sections must be fabricated from plate. This will add to the cost of the contract. It also means that a designer planning a weathering steel AESS project must vision in terms of custom plate type details.

Few steel service centers will stock a larger inventory of weathering steel, given its specific use for bridge applications. The material would be ordered from the mill on request and, unfortunately, typical orders such as for an exterior wall element represent low tonnage for a service center.

Connection methods for weathering steel are similar to other types of steel as bolting and welding are common. The weld material is similar in chemical make-up to the weathering steel so that it will weather consistently with the adjacent steel.[1] This applies also to bolts that are manufactured from a weathering grade steel.

Weathering steel is also available in thinner sheets for roofing and cladding. It must be kept free from debris such as leaves, pine needles, etc. These waste products retard the wet/dry cycle necessary for weathering steel, thereby accelerating corrosion. Also, in an accelerating environment, loss of material may be more significant and is liable to cause perforation of very thin sheets.

Weathering steel would require the use of the Custom AESS category to include specialized instructions to ensure a quality weathering steel product.

Although not structural proper, weathering steel is finding uses in landscape design, like in this barrier in Wellington, New Zealand. The plate thicknesses tend to be minimized to save cost. This can be problematic if exposed to excessive wet/dry cycles or if natural materials are allowed to build up on the steel. Thinner material will eventually corrode.

This parking garage in Auckland, New Zealand, is using a weathering steel plate for its exterior cladding. The plate is being installed over a painted HSS framework to add rigidity to the plate and make it appear more substantial in thickness.

The plate is being simply attached to the HSS framework using metal fasteners through the front of the material. This method of attachment requires careful coordination of the placement and patterning of the pre-drilled holes in the plates and frames.

STRATFORD TOWN CENTRE LINK | LONDON, ENGLAND

Knight Architects with Buro Happold, 2012

The Stratford Town Centre Link was designed and constructed as a major gateway to the 2012 London Olympic Games. It is noted as being the first live launch of a bridge over an operational station in the United Kingdom. The project made extensive use of weathering steel, with all elements custom-fabricated from steel plate. Welding was used as the predominant connection type. The design comprises a ladder beam arrangement with steel side trusses, fabricated entirely from weathering steel and following a Vierendeel type forming two equal spans of 65m/213ft. Glazing is used within the framed openings of the trusses to provide fall protection over the tracks. At 130m/427ft in length it is thought to be the longest pedestrian bridge constructed entirely of weathering steel plate.

The 12m/39ft-wide bridge is subtly curved in plan. The trusses vary gradually in height from 4.5m/14.7ft at the ends to 6.5m/21.3ft at the center pier. The outermost truss develops a deliberate lean of up to 15°.

The Stratford Town Centre Link connects Stratford Station to the Westfield Mall that was the entry point for the 2012 London Olympic Games. The bridge is constructed as a slightly arched truss whose sides tilt outwards. The bridge makes use of all-welded connections in keeping with its form. The trusses of the bridge were designed with built-in drip details and chamfered plates to prevent pigeon roosting.

The welds have been left "as is" to create the technical aesthetic of the bridge.

The bridge is curved in plan as well as elevation. The outer profile of the vertical and horizontal members is shaped to create sharp edges and chamfered planes that are defined by light and shadow to accentuate the form and reduce the visual mass in elevation.

THE GLASS BRIDGE | PERTH, AUSTRALIA

Donaldson + Warn; Capital House Engineers; artists David Jones, Kevin Draper and Richard Walley; John Holland Constructions, 2003

The Glass Bridge in King's Park makes extensive use of many types and forms of weathering steel. From cast panels for the decking, to the pylons and railings, weathering steel has been selected to help the bridge fit into the landscape of the park while providing a durable and low-maintenance option. The project was conceived as a large element of public art whose footprint in the natural environment had to have minimal impact on the forest floor.

Glass railings contrast with their stainless steel fittings. The cast floor elements provide a more slip-resistant surface.

A view to the underside of the bridge showing the support system. Wide use of welding was necessary to custom-fabricate the elements from plate steel.[2]

The detailing of the handrails makes use of welded plate as other forms of this material such as hot-rolled sections are not available.

Stainless Steel

Significantly more expensive than regular carbon steel, stainless steel is the most corrosion-resistant form of steel that can be selected for an architectural project. Stainless steel offers the advantage of corrosion protection as an integral feature of the structural member. It has a remarkable finish to itself and for these reasons requires far less maintenance than regular steel protected through methods of galvanizing, metalizing or with painted or intumescent coatings.

There are many different grades of stainless steel as a function of its chemical composition. The most common grade for architectural steel would be 304. However, if the environment has a high level of atmospheric pollution or the location is marine, then 316 should be selected for its superior corrosion-resistant qualities.

There are significant cost premiums for stainless steel even beyond the cost of the material. From a structural perspective, it requires a different set of calculations as its behavior is very different from regular carbon steel. Only dedicated tools must be used to cut and finish the steel, as tools used also with carbon steel would embed small particles of carbon steel in the stainless steel, causing rust spots to occur.

The range of material sizes is not as wide in stainless steel as for regular carbon steel. Stainless steel would require the use of the Custom AESS category with specialized instructions.

The exterior domes in Robson Square, Vancouver, BC, Canada, were replaced with stainless-steel glazed structures. The damp exterior environment is worth the extra expense to provide such a corrosion-resistant, low-maintenance material. The beauty of the finish in stainless steel requires that extra care be taken in welding and detailing.

The domes use a combination of circular HSS and plate material. The welds are neatly done and left unremediated as grinding can cause problems when dealing with stainless steel. The major elements of the domes were prefabricated in the shop of George Third & Son, where conditions are optimal. Site welding was not desired. Special stainless-steel bolted connections blend in very subtly with the other welded connections. Intermittent welds can be seen where a stainless plate contacts the glass. Continuous welds would have caused heat deformation to the material and the intermittent welds are barely visible. In the case of stainless steel it is not possible to use filler to provide a continuous weld appearance.

FIRE PROTECTION SYSTEMS

Steel, although not combustible, suffers sudden failure when exposed to heat and must be protected from exposure to fire. The primary concern for fire protection is to allow the safe evacuation of the occupants. The secondary desire is to further maintain the structural integrity of the building so that repair is possible. In many situations, the steel will need to achieve performance levels associated to more than one of the AESS categories at the same time.

Analysis of fire-related code issues at the outset of the project ascertains the level of exposure permitted. This determines the approach to the types of finishes possible or required for the project. In certain high-risk uses, steel must be protected from exposure to the extent of precluding any exposure. Between the extremes of concealment at the highest level of protection, to a simple paint finish for very low-risk situations, most projects will use an intumescent coating. Spray Fire-resistive Materials (SFRM) are used less frequently, as they tend to obscure details and only leave the essence of the form expressed.

Intumescent Coatings

Intumescent coatings simultaneously provide a fire resistance rating and a painted appearance for exposed steel. They contain a resin system "pigmented" with various intumescent ingredients that, under the influence of heat, react together to produce an insulating foam or "char". This char layer has low thermal conductivity as well as a volume that is many times larger than that of the original coating. The char layer reduces the rate of heating experienced by the steel, extending its structural capacity and allowing for safe evacuation. As this material can extend the fire resistance rating of exposed steel to a maximum of two hours, it has become quite popular for use with AESS applications. The fire resistance rating is in part dependent on the type and thickness of the coating as well as the type of fire that might be anticipated in the building use. Increasing the fire resistance rating is usually achieved by applying multiple coatings of the product.

The required thickness of the coating is determined by the thickness of the structural steel member. Thin or light members will require more coats than heavier members. It is sometimes more cost-efficient to increase the thickness of the steel, as this can decrease the number or thickness of the intumescent coatings: the increased cost of steel is significantly less than the extra cost to increase the thickness of the intumescent material. Structural steel is inherently a more sustainable material, so the reduction of the amount of coatings is preferable.

Intumescent fire protection application is preceded by the application of an approved primer. Not all primers can be used and an unsuitable primer may interfere with the application of the intumescent coating system.

Traditionally, intumescent coatings have been applied onsite to steel structures during the construction phase of the building. In-shop application has become a more common practice as better control of application conditions can be achieved. Shop applications can provide for the controlled venting needed for solvent-based systems. Shop conditions can also provide better control of temperature and relative humidity and hence further better drying. Controlled drying in the shop means better finish and also the coated steel sections cannot be moved until they are hard enough to resist damage. These members must be more carefully handled during transportation and erection as any damage must be properly repaired in order to preserve the integrity of the fire protection system.

Intumescent coatings are either acrylic or epoxy-based. Acrylic coatings can be either water or solvent-based. The acrylic coatings are field-applied. The water-based material is ecologically friendly but takes somewhat longer to dry and is mostly used for interior applications. Water-based coatings are typically applied when the relative humidity is between 40% and 60%. If there is a concern about the presence of high VOCs on the project, a water-based product can be used if the humidity levels are kept low. It is important that the layers are allowed to thoroughly dry between coatings. Water-based products take longer to dry where humidity levels are high and temperatures are low.

The solvent-based coating is more robust. Epoxy coatings are normally shop-applied and can be used on interior or exterior structures. They are more durable than acrylic coatings and can also be used to provide corrosion protection. Where access for finishing may be an issue, shop-applied epoxy coatings may offer savings. Solvent-based fireproof coatings can be applied with relative humidity of up to 85%. Solvent-based products can dry faster but can also strike back to dissolve prior layers if the drying time between layers is insufficient.

The intumescent coating system can include a "top coat". This provides a hard protective coating to a surface otherwise prone to fingerprinting and difficult to clean. It is important to note that white or light colors will tend to yellow with time. If color matching is an issue, this behavior should be taken into account when mixing intumescent and painted finishes in a project. No exact color matches are possible when combining intumescent and regular paint-finished steel. The nature of the intumescent finish alters the color of the coating with a slight change in hue or tone. In any case, the product must be permitted to completely cure prior to the application of any final top coats. If not allowed to cure properly, exposure to heat (sun) leads to blistering of the uncured solvents as they attempt to expand and escape.

There are thin and thick intumescent coating systems. A thin coating comprises thicknesses of 0.5-0.6 mm/0.020-0.024in, thick coatings thicknesses of up to 13mm/0.5in. Because the wet film needs to be relatively thick (several hundreds of microns according to the particular formulation), intumescent coatings are often thick to avoid slumping and runs while wet. Several coats may need to be applied to build up to a total dry coat thickness that is sufficient to fulfil the requirements for the intended fire protection.

Intumescent coating is used to allow the exposure of the steel in the atrium of the Bloomberg Building in New York City, NY, USA, designed by Pelli Clarke Pelli Architects.

The detailing of the Bloomberg Building has been designed with the thickness and finish of the coating in mind. The details are simple and the reveals around the connections are designed to add evidence of form to the connections. The texture of the finish is evident. There would be no benefit in contouring and blending the welds.

Capital Gate in Abu Dhabi, UAE, designed by RMJM Architects, uses an intumescent fire protection strategy to allow full exposure of its custom-fabricated diagrid frame.

The nature of the intumescent coating has been considered when designing the steel. The thickness supports the blending of the connected sections. Both sharp corners and fine details have been avoided.

Although intumescent coatings convey the appearance of a painted finish, the texture is not the same. Thin coat systems create a finish that resembles an orange peel. The thicker system has enough substance to conceal some of the finer details that might go into the design of the AESS connections. If badly applied, a thick system can give a very uneven, textured appearance. For these reasons intumescent coatings are not always deemed to be the best solution, although they allow the use of exposed steel in an increasing number of occupancies. On the other hand, the use of intumescent coating often precludes the need for fine finishing, as it is thick enough to cover up surface imperfections that would be unacceptable if a standard paint finish were employed. If a very smooth, high-gloss finish is desired, this system requires additional surface treatment.

Care must be taken when using thick coatings in high-traffic areas or where they can be subject to vandalism. Damaged intumescent coating must be properly repaired to maintain the required fire resistance rating. Color matching can then be an issue.

The diagrid HSS columns at the World Financial Center entry pavilion in New York City, NY, USA, designed by Pelli Clarke Pelli, are having the final touches rolled onto the intumescent coating. This structure required extensive site welding and arrived to site with a primed finish. The site application of the coating system to the final required quality was very time-consuming.

The fire protection on The Leadenhall Building in London, England, designed by Rogers Stirk Harbour + Partners, was achieved through a shop-applied marine-standard epoxy-intumescent coating on the exposed steel frame. The thickness varies from 3mm/0.125in to 12mm/0.5in as a function of the thickness of the steel. Thinner steel requires thicker protection to achieve the same rating. The rating achieved is 90 minutes. Epoxy-intumescent coatings are normally shop-applied and this necessarily infers more careful handling during transportation and erection. However, some damage to the finish being unavoidable, a permanent artist touch-up person was on site to remediate damage as it arose.

Sprayed Fire-resistive Materials (SFRM)

Although not the usual case for AESS installations, cementitious or fibrous fire protection may be an option if the steel is located at a distance from view or touch, as in the case of AESS categories 1 or 2. If such a finish is to be applied, there need not be the same level of surface preparation. Such projects would fall into AESS 1 to avoid wasted time and expense on the fulfilment of requirements that do not apply here.

There are various types of SFRM, ranging from a lighter spray-applied material to a more dense material that is trowel-applied and can have a top coat. The thickness of any variety will mask most design details. Also, this material will be significantly more difficult to keep clean, so its use should be aligned with the use of the space and the presence of particulate matter – it can be problematic in train, bus and subway stations, for example.

The shape of the sections is another factor when considering a trowel-applied SFRM. It is very difficult to achieve a high-quality SFRM finish on curved surfaces.

The major public areas of the Experience Music Project in Seattle, WA, USA, designed by Frank Gehry, are permitted by code to use intumescent-coated steel with a fire suppression system.

Some areas require the use of a high-density fireproof coating as well as a fire suppression system.[3] The nature of the steel elements, curvature and dim lighting helps to make this variation in finish recede from notice. The detailing of the steel recognizes the impact of the protection system and is kept quite simple, leaving it to the overall form to provide the major aesthetic expression.

The high-density fireproofing masks the detailing of the curved structure, while the overall form remains strongly evident. The muted tone of the selected color helps the fireproofing to visually blend in with the intumescent-protected exposed structure.

Concrete-filled Steel Tubes

Concrete has long been used as both a fire protection material as well as a composite material for structural reinforcement. In addition to its use to surround the steel, creating a depth of cover between the fire source and the steel, it can be used to fill HSS tubes, thereby increasing their resistance to fire. Filling provides less resistance than encasement. Depending on the required time of fire resistance, the use of concrete-filled steel tubes may need additional protection by an intumescent coating. Concrete-filled tubes are very suitable in AESS applications for larger structural member sizes as a means to allow exposure of the steel.

The large delta frames (left) that create the unusual structural support system for QRC West in Toronto, ON, Canada, designed by &Co Architects, uses concrete-filled steel tubes to meet both structural load capacity and fire resistance requirements.

The concrete will be pumped into the base of the legs (top) until it reaches the top of the structure. This process ensures that the tubes are completely filled. The inlets will be removed and the surfaces remediated prior to the subsequent application of intumescent coating.

A concrete-filled steel tube column system offers many advantages compared with standard structural steel or plain reinforced concrete systems for projects where expression of the steel is the desired aesthetic:

Better interaction between the steel tube and the concrete: Local buckling of the steel tube and strength deterioration are delayed by the restraining effect of the concrete. Vice versa, the strength of the concrete is increased due to the confining effect provided by the steel tube. Concrete spalling is also prevented by the tube and creep of the concrete is smaller than in ordinary reinforced concrete.
Cross-sectional properties: The steel ratio in the concrete-filled tube cross-section is much larger than in reinforced concrete and concrete-encased steel cross-sections.
Construction efficiency: The labor for forms and reinforcing bars is omitted and the concrete casting is done by tremie tube or pump-up methods. This enables a cleaner construction site and a reduction in labor, construction cost and project length.
Fire resistance: Concrete improves the overall fire resistance, so that fireproof material such as additional intumescent coatings can be reduced or omitted.

Cost performance: Because of the merits listed, better cost performance is obtained.
Environmental impact: The environmental impact can be reduced by omitting the formwork, specifying steel with a high recycled content and using high-quality concrete with recycled aggregates. There is a certain elegance achievable when designing with concrete-filled steel tubes that is not easily matched by using other types of steel sections, with or without a fire-resistant covering or casing.

These very large tubular columns at the Guangzhou IFC in Guangzhou, China, designed by Wilkinson Eyre Architects, are filled with concrete as part of their fire resistance. They also have a trowel-applied coating to complete the assembly to meet the required rating for this building, which belongs to the class of Supertall Towers. An additional top coat has been applied to provide a durable, glossy finish suited to the high-end hotel function.

5 D Saída
Saída 6 D

MEMBER CHOICES

CHAPTER 6

THE TEXTURE OF AN AESS PROJECT

The combination of the exposure of the steel, the selection of member types and expressed connection details creates the structural texture of an AESS project. This chapter examines the role that the basic choice of member types will have on the development of the aesthetic expression of a project, its relative cost as well as the type of connection detailing that will naturally ensue. There is a relationship between member types and connections that tends to drive the detailing in many projects. It is difficult to speak of one without the other and there are multiple ways to approach the selection of member types, connection types and AESS categories. Certain types of structural members tend to support the detailing choices aligned with the specific categories of AESS.

Oriente Station in Lisbon, Portugal, designed by Santiago Calatrava, constructed for Expo '98, stands as a project that influenced the development of AESS. The detailing of the fabrication made predominant use of custom-fabricated sections, a practice that has been reduced in deference to using a wider range of standard structural and hollow shapes.

AESS structures, by their inherently exposed nature, put a greater than normal emphasis on connection design. The detailing language of the connections must feed into the overall aesthetic desired for the structure, as well as be constructable and within reason to erect. Although the details that are normally used for AESS structures include some fairly standard connection methods, these are mostly modified as a way of enhancing the architectural expression of the structure, and by their nature are likely to present challenges for both fabrication and erection as discussed in the previous chapters.

The proper application of the Category System approach to specifying and detailing an AESS project can assist in developing a layered or tiered approach to establishing the appearance of the steel, putting emphasis and budget towards elements that will have the most impact and allowing for simpler detailing on elements that are distant to view.

CHOOSING A MEMBER TYPE

The choice of the type of structural section greatly impacts the potential expression in the connection design. It also tends to form an alignment with higher and lower AESS categories as the type of member will either require or infer aspects of connection design. Almost all section and member types can be joined using bolting or welding. The exception would be tension systems, as the extreme slenderness of the rod and cable members requires the use of specialty connections to interface with the end conditions of the members.

Both hot-rolled steel sections and even more so HSS will have a rounded appearance due to their manufacturing processes. For hot-rolled structural shapes the rounding occurs at the interior connections of web to flange of the member and for hollow structural shapes the outside corners are rounded. The thicker the plate from which the section is rolled, the more pronounced the curvature.

For higher AESS categories there is a tendency to prefer sharper, crisper edges. This often leads to the use of custom sections that are fabricated from specialty cut plate. The appearance might be that of a standard hot-rolled or hollow section, but the reality is far different. There is a significant cost premium to custom-fabricate the members in addition to that of the fabrication of the connections. Conversely, there are economies to be gained if standard shapes can be used in innovative ways.

In this way the simple choice of the types of structural members for the AESS system will imbue a project with a certain texture or feel. And although much of the cost of fabrication is a result of the complexity of the fabrication of the connections, the choice of the members contributes to this cost as a function of:

- the amount of remediation or surface preparation required for the member itself (hence the suggestion that grinding the weld seam on an HSS member might be unnecessary and better handled by simple orientation or by the increased fabrication with custom members from plate)
- the types of connections that seem to naturally align with the choice of member
- the choice to specify premium or difficult-to-obtain structural types (like oversized tubes, tapered tubes or elliptical hollow sections)
- the choice to custom-fabricate the members from plate, which requires additional remediation pushing towards AESS 3 standard as a minimum
- the requirement to bend members
- the need to resist compression, tension, bending, eccentric loads or combined loading

There is nothing inherently right or wrong when choosing a specific AESS member. All types of compression members can be sized and detailed to satisfy loading requirements. The clear identification of tension-only elements in a system can permit the selection of tension-specific materials and the associated detailing, taking into account the inherent textural differences between cables and rods.

HOT-ROLLED STEEL SECTIONS

Hot-rolled steel shapes have a long history of use in expressed steel structures as they predate the invention of HSS in the 1970s. Much of the exposed detailing created by Ludwig Mies van der Rohe, Philip Johnson, Charles and Ray Eames and the Brutalists was based in the availability of this type of steel.

Crown Hall at the Illinois Institute of Technology in Chicago, IL, USA, built 1954–1956 (left), stands as a seminal example of the work of Mies van der Rohe and the use of orthogonally-based hot-rolled steel materials.

Mies' Farnsworth House in Plano, IL, USA, built in 1951 (right), makes predominant use of standard sections.

Standard structural shapes – W (Universal), C and L – are used in a wide range of AESS applications. The detailing of these shapes can present greater challenges in the creation of smooth or artful connections. One of the main issues with the use of standard shapes is the orientation of the members to allow for drainage (if placed outdoors) and also to prevent the accumulation of dirt or debris in the hollows. Issues pertaining to thermal bridging, heat loss and durability have changed the way that contemporary AESS is detailed for exterior expression.

Although many of the projects that choose hot-rolled shapes tend to use AESS 1 or AESS 2 standards, this does not necessarily translate into an identical or even equal approach to the aesthetics of the projects. There is a significant range of appearances when using fairly standard members. The overall look is greatly influenced by the choice of geometry and connection detailing, connections being made using welding, bolting, gusset plates, angles and pin connections.

Seldom do projects limit the use of sections to one type or another. It is normal to use a variety of member types and select those that are best suited to the job for which they have been chosen. It is more common to see wide-flange (Universal) sections selected for columns, beams and truss members and less common for use as tension or cross-bracing, where lighter sections are preferred and the end connections more artfully handled.

The J. R. Thompson Center in Chicago, IL, USA, designed by Helmut Jahn in the mid-1980s, reflects an early application of AESS. The use of color evidences its alignment also with Postmodernism. The height of the atrium allowed for a use of steel differentiated by viewing distance.

The project makes widespread use of a combination of standard structural shapes, including wide-flange (Universal) sections and angles for the truss support system for the roof of the glazed atrium, as well as HSS for the walkway support system. The segregated use of shapes to align with aesthetics, functions and distance of view is clear. The connection details are very expressive of the method of connection and the project makes little use of expensive remediation methods.

The exposed steel on the exterior structure of the National Works Yard in Vancouver, BC, Canada, designed by Omicron, uses standard wide-flange (Universal) sections to create most of its frame. This gives a more technical feel to the steel, which has been combined with engineered timber and exposed galvanized steel decking. This could be classed as AESS 2, given the use of standard sections and lack of weld remediation.

The connection detailing of the National Works Yard demonstrates the combined use of welded and bolted connections. The bolted connections are used to assist in on-site erection and are made visible as a part of the design detailing by expressing the bolts themselves. They give the structure a more technical feel that suits the function of the building.

The support system for this exterior walkway in Auckland, New Zealand, has been fabricated from wide-flange (Universal) sections. Connections have used a combination of welding and bolting as a function of the in-shop assembly of larger components and on-site erection requirements. Standard industrial cross-bracing has been used to stabilize the frame. This project could be classed as AESS 2.

Standard steel sections lend themselves to fairly typical connection detailing in the Auckland walkway. The detailing, when carried out according to AESS standards, succeeds in elevating the use of the steel to architectural qualities. Here some typical bracing elements have been added for stability, lending a finer texture to the overall appearance of the structure.

The technical or industrial appearance of wide-flange (Universal) sections is a good match to galvanizing as a final and corrosion coating, as has been done in the Calgary Water Building in Calgary, AB, Canada, designed by Manasc Isaac Architects. This project makes predominant use of shop-welded connections within the truss, while bolted connections are used between larger elements to simplify the erection process.

Wide-flange (Universal) members have been selected to create this exposed steel truss at the Gaylord Conference Center in Dallas, TX, USA. The member-to-member connections that form the transfer nodes of the truss are using simple bolted gusset plates. The truss could be classed as AESS 1 given the nature of its detailing. The use of castellated beams adds textural interest to the system without requiring extra effort to be spent on connections.

The temporary stands used for the Aquatics Centre for the 2012 Summer Olympics in London, England, designed by Zaha Hadid, made use of a combination of wide-flange (Universal) members, angles and HSS. This type of member choice is quite common in sports facilities and outdoor stadiums.

The bolted detailing was chosen to facilitate the deconstruction and removal of the extra viewing stands immediately following the event in progress in this image.

Wide-flange (Universal) sections were used in the creation of the vast steel diagrid structures for the Seattle Public Library in Seattle, WA, USA, designed by Rem Koolhaas. Larger sections were shop-prefabricated using welding and bolted together on site. It would have drastically escalated the cost to fabricate this steel if the complete customization of the members had been required. The connection between the adjacent flanges of the members required a kind of weld remediation that would class this at AESS 3.

The steel framing in the access zones for Heathrow Airport Terminal 5, in London, England, designed by Rogers Stirk Harbour + Partners, Arup and Watson Steel Structures Limited, uses a combination of hot-rolled sections and various sizes of HSS to create the aesthetic of the brightly painted exposed structure. This is in great contrast to the large white support system that spans the vast departures hall. What elevates the interest in the portion depicted, which makes predominant use of standard hot-rolled sections, is the nature of the pin-connected details and the simple alignment of the web stiffeners at the connections. As this structure is viewable at fairly close range it could be classed as AESS 3, having benefitted from remediation efforts.

TUBULAR OR HOLLOW STRUCTURAL SECTIONS (HSS)

The emergence of High Tech architecture in the 1970s aligned with the invention of hollow structural sections. Although connection detailing has changed and become more subdued since that time, these types of members are more often used in AESS projects than hot-rolled shapes as HSS are generally perceived as more elegant and less technical in appearance. The connections associated with HSS already deviate greatly from the standard structural connection details that were developed for standard shapes. These were addressed at length in *Understanding Steel Design: An Architectural Design Manual* in "Chapter 3: AESS: Its History and Development". For many designers this makes HSS the "go to" structural material when conceiving of AESS projects, even though this need not be the case.

Therefore, tubular steel – either HSS or mechanical pipe – is often chosen when creating AESS projects. Standard hollow structural steel shapes are circular, square or rectangular. Mechanical pipe is produced round. The choice of the cross-sectional shape will have a tremendous impact on the design and appearance of the connections. The geometry of the connection – planar, simple angle or multi-member intersection – will impact the cost and complexity of resolving multiple HSS shapes. In some instances the joint can be resolved by cutting and welding. In other instances plates may be needed to simplify the intersection, although contemporary methods of computer-assisted cutting are making such work easier.

HSS are actually created from flat plate. Large rolls of flat plate are uncoiled and worked through a series of progressive roller/dies to form the plate into a tube, which is welded along its length to form a default circular section. The tube can then be fed through a series of rollers that reshape it into square or rectangular sections. The section sizes are a product of the initial size of the coiled plate material. This is referred to as the Electric Resistance Welding (ERW) process. Alternatively, the Form-Square Weld-Square process can be used to bypass the creation of the circular cross-section and directly form orthogonal materials. The method of choice will vary by manufacturer but can be helpful to know when sourcing material.

The linear weld seam is visible on the exterior of the section. Details must acknowledge this and either orient the seam away from view or have the seam ground smooth if near to view. The seam may not be in a consistent location from supplier to supplier and will vary in position as a function of the size of the HSS shape. Tubular sections greater than standard fall into the Custom AESS category as they must be manufactured from plate material using a brake press or by using a plate rolling machine. For more detailed information refer to *Understanding Steel Design: An Architectural Design Manual*, "Chapter 8: Curved Steel".

While HSS is normally produced using a welding process, pipes tend to be the result of an extrusion process. Therefore the shape of mechanical pipe is seamless. When designing with HSS, the AESS characteristics require the designer to look at the orientation of the weld seam in the design. A welded seam will tend to be visible even after grinding, even after finish coatings are applied, although depending on the coating process used, as one can only grind perpendicular to a surface. Therefore, it is preferable and less expensive to simply orient this natural occurrence consistently or away from the dominant angle of view.

Round mechanical pipe is sometimes chosen for complex trusses as it does not have a seam and hence presents one less issue of concern for the final appearance of the structure. However, it must be recognized that the structural and material characteristics of pipe are quite different from HSS and the two cannot simply be interchanged. It is necessary to have a full team consult to make this decision. Pipe is also not permitted in seismic design due to its different structural characteristics.

The maximum diameter that can be mass-manufactured is approximately 1,100mm/42in at the time of writing. There will be variability in the availability of different section sizes, but this is not the case for different diameter ranges. For large quantities (i.e. over 50 to 70 tonnes) an order can be placed directly to the structural tubing mill. HSS with a diameter greater than 400mm/16in generally require special ordering. In most cases big tubes will be custom-manufactured and will require a minimum 100-tonne quantity when ordering. As helical welds are sometimes proposed for large HSS sections, it is important to be sure to discuss this with your fabricator and explicitly exclude these in your AESS specification documents if they are not acceptable.

The Humber River Bridge in Toronto, ON, Canada, designed by Montgomery and Sisam Architects, uses large-diameter, helically welded tubes for the main spanning sections of the arch.

This close view of the bridge arch shows the helical weld. The scale of the member and its use were sufficient to result in a decision not to remediate the welds. This would have added significantly to the contract and not have been of adequate benefit.

Tapered tubes are not a regular manufactured product. They must be custom-fabricated from a trapezoidal plate that is rolled to form a tapered pole and the seam must be welded. These shapes fall into custom fabrication (see page 113).

The ultimate texture or aesthetic of a tubular AESS project will greatly depend on the method of attachment, i.e. the choice between welded or bolted connections. This will be discussed in greater detail in the next chapter, but it is important to note, when examining the projects in this chapter, the impact that the connection type has on the overall feeling of the structural system.

Round Hollow Structural Sections

Round HSS is often chosen for projects that desire a sleeker, controlled appearance. There is an inherent smoothness to the members that lends well to welded connections. Modern cutting methods make the precision cutting of round-to-round connections more straightforward. The wide range of connection details for tubular products will be examined in the following chapter. The range of applications illustrated here is intended to demonstrate the relative textural appearance of the steel as a function of the member shape, selected finish and general nature of the connections.

Torre Vasco da Gama, built for Expo '98 held in Lisbon, Portugal, designed by Leonor Janeiro, Nick Jacobs and SOM, uses round hollow members for its truss-like structure. Fully welded round tubes are often chosen for exterior structures for serviceability reasons. The round shape is good at shedding water and the surfaces tend not to collect dirt and debris. Welded connections seal the system and also provide for easier cleaning and maintenance.

In addition to round tubes, the project makes use of custom-tapered sections (top) to gently step down the sizes in accordance with changes in loading requirements. This allows the structure to decrease in diameter towards the top and become lighter in appearance. Tapered tubes are custom-fabricated from plate (see page 113). The curvature is achieved through a combination of curved sections and straight segments that have been welded together. The large scale of the curve allows for this method, while tighter curves would reveal the faceting of the form. The exposed floor system for the viewing platform uses standard wide-flange (Universal) sections with profiled decking.

The texture of the tubular frame structure that supports the massive sun-shading device on Capital Gate in Abu Dhabi, UAE, designed by RMJM Architects, has a more technical feel due to the use of bolted connections that rely on the insertion of a welded plate into the section.

The trusses that provide wind bracing and lateral support for the large glazed façades at the Beijing Capital International Airport in Beijing, China, designed by Foster + Partners with Arup, are fabricated from round HSS material. The choice of welded connections within the trusses provides a very smooth appearance. The tubes that form the exterior members of the truss are larger in diameter than the cross members, making the welding of the smaller element to the larger one easier to fit.

The Phoenix International Media Center in Beijing, China, designed by BIAD UFo, exploits the ability of round tubular steel for its ability to easily respond to the creation of curves in this eccentrically curved structure.

The addition to King's Cross Station in London, England, designed by John McAslan and Partners with Arup, makes predominant use of round tubular material to create this dynamic diagrid-like structure. Round tubes were selected for their ease of use in situations requiring curvature of the structure.

Square or Rectangular Hollow Structural Sections

Square or rectangular HSS tend to engender a more technical look, although less industrial than would wide-flange (Universal) sections. HSS in general can be finished to an AESS 4 standard if desired, although AESS 2 or 3 would be more usual. This type of HSS supports a choice of welded or bolted connections. Again the choice of connection type plays into the overall aesthetic or texture.

A simple frame of square HSS columns and beams supports the large canopy in front of the Melbourne Museum in Melbourne, Australia, designed by Denton Corker Marshall. The material and detailing support the intended simplicity of the design.

The lightness of the HSS frame accentuates the enormity of the cantilevered plane that it seems to mysteriously support. A light steel mesh on the underside of the plane allows for a view through to the rather substantial braced structure that keeps it rigid.

The square HSS of the Canadian War Museum in Ottawa, ON, Canada, designed by Raymond Moriyama Architect, were designed to provide the aesthetic of a war-torn landscape. The erratic twist to the shape of the space is complemented by an expressive sequence of bolted connections between the members, while the internal connections of the trusses are fully shop-welded.

Galvanized coating in combination with expressed bolted connections, as well as the chaotic layout of the structure govern the texture and expression of the AESS at Federation Square in Melbourne, Australia, designed by Lab Architecture Studio with Bates Smart Architects.

Glazed façades such as the one used at the Wisconsin Intermodal Terminal in Milwaukee, WI, USA, designed by Eppstein Uhen Architects, can allow the AESS system to be expressed to the exterior without having to expose the system to the effects of weathering. This relieves pressure on the choice of coatings and detailing to achieve higher durability levels.

The choice of round HSS for the bracing that crosses over the passenger waiting area (top right) alters the feeling and texture of the space. The simple pin connections at the ends of the tubular braces contrast with the fully welded connections between the rectangular sections.

The structural system uses square and rectangular HSS with fully welded connections (right). The texture of this structure results from a unique combination of this section type with an irregular diagonal pattern. To achieve this look the connections between intersecting sections must be fully ground and blended so that the connections become close to invisible. This would put the structure at AESS 4 in detailing, with the odd geometries adding to the challenge of fabrication.

ELLIPTICAL HOLLOW SECTIONS (EHS)

Elliptical hollow section tubes are relatively new to the AESS scene. Their use started in Europe and they are slowly making their way into projects all over the world. Elliptical hollow sections have greater bending capacity than circular HSS of a similar dimension or weight due to disposing of strong and weak axis directions, while maintaining a smooth closed shape. There is also reduced visual intrusion compared to regular circular HSS if a member is viewed from one predominant direction.

All elliptical hollow sections are produced as hot-finished hollow structurals, with major-to-minor axis dimensions of 2:1. They are produced as continuously welded sections, joined by high-frequency induction welding and finished to their final shape at extremely high (normalizing) temperatures, with the outside weld bead removed but the inside weld bead typically left in place. As elliptical hollow sections are normally used in high-end AESS projects, the removal of the weld seam was deemed important for aesthetic performance.

Due to the hot-finishing process elliptical hollow sections dispose of a fine-grain structure, uniform mechanical properties, excellent weldability and negligible residual stress; they are suitable for hot-dip galvanizing and are applicable for dynamic loading situations. Elliptical tubes have similar material properties as regular HSS and similar connection methods can be used in the connection detailing of the members.

Elliptical tubes are gaining popularity as a glazing support system because the tubes can be oriented to minimize their visual impact in keeping with the aesthetic aspirations of structural glazing systems. The tubes are normally placed with their narrow edge facing, which is the best orientation to resist horizontal wind loading. Placing the tubes on their flattest side is not good to prevent deflection from self-load when support distances are great, and if installed flat need to be braced back to the primary structure or given additional support by employing a hanger system.

Sloped elliptical tubes are used at the Stratford DLR Station in London, England, designed by SMC Alsop Architects, to provide architectural interest to the simple structure of the station. The end connections are hidden to focus attention on the shape of the members, which are also highlighted by the use of bright colors.

Neo Bankside Housing in London, England, designed by Rogers Stirk Harbour + Partners, uses elliptical tubes to create its external bracing system. Much effort has been made to develop a system of details to connect the elements. They are oriented with their wider face to view, thereby making them feel wider yet lower in profile to the skin of the building.

The Y-shaped structural supports in Barajas Airport Terminal 4 in Madrid, Spain, designed by Rogers Stirk Harbour + Partners, uses elliptical tubes that have been bent and attached using plate steel. The narrower profile of the ellipse creates a very smooth-looking support system.

Heathrow Airport Terminal 5 uses elliptical tubes behind the multistorey glazed façade to provide bracing. The profiles are oriented to minimize the apparent size of the members in relation to the horizontal mullions of the glass.

The placement of the horizontal shading devices on the exterior of the glazing at Heathrow links back to the elliptical tubes. The exterior shades work to mask the location of the tubes when viewed from outside. HSS struts provide bracing of the tubes back to the primary structure.

Given the temperate climate of Porto, Portugal, sufficient lateral/wind loading support could be provided to the glazing of the Porto International Airport through the exterior placement of horizontal elliptical tubes. The tubes are tied back to the primary round HSS supports as well as being hung with stainless steel rods to limit deflection.

CUSTOM-FABRICATED SECTIONS

The complete custom fabrication of members tends to be reserved for AESS 3 and AESS 4 categories due to the cost premiums required for more extensive remediation associated with higher levels of welding and increased aesthetic expectations. Plate steel is often selected when standard members do not fit with the desired aesthetic or the structural capacity cannot be met with standard section sizes. Where plate material is fabricated into orthogonal shapes, it provides the crisp corners that may be desired. Plate material can be fabricated into a wide range of tapered or irregular shapes, including curved forms. The cost premiums can be extremely high as they comprise additional fabrication costs (labor and time) for all members, where in most other systems such cost premiums are predominantly associated with the connections.

There is a very wide range of use of plate material in the fabrication of AESS elements – from creating the members, complex connections or nodes, to larger elements that combine aspects of a column and a spanning-type member. In particular, custom sections will be required for extra-tall column elements whose structural and aesthetic requirements are not satisfied by standard wide-flange (Universal) or hollow members.

The (Siemens) Crystal in London, England, designed by Wilkinson Eyre Architects, has adopted an overall geometry that would suggest the use of custom plate sections.

The extremely tall volume of the interior exhibition space of The Crystal placed particular requirements on the design of the columns, which could not be met using standard materials.

The designers of the Crystal chose to use plate material to create the triangular and trapezoidal columns and beams. These have been detailed to accentuate the meeting of the planes, allowing the overlap and extension of the material that creates a distinct shadow line to the edges of the elements. The ensuing faceted look, in keeping with the overall design aesthetic, obviates the need for excessive weld remediation. The layout of the beams was coordinated with the spanning capacities of the exposed steel decking to eliminate the need for an intermediary support system. In this way it was possible to better showcase the primary structure.

The multi-storey column at the Lloyd D. George U.S. Courthouse and Federal Building in Las Vegas, NV, USA, designed by Cannon Design, required custom fabrication due to its extreme height and its feature role in the overall design.

The column of the Lloyd D. George Courthouse and Federal Building is comprised of a number of custom segments with a round HSS at the core. The X-shaped support picks up the loads from the trellis in a two-way spanning arrangement. Even though the trellis beams actually have a dominant span direction, the appearance of load symmetry is achieved by ensuring that the bottom flanges of the beams align perfectly, giving the illusion of equality. The beams that span between have been slightly overdesigned to achieve this appearance.

The multi-storey columns in the public atrium of the Dalian Conference Center in Dalian, China (left), use plate steel for the custom sections. These have been enhanced with pairs of ribs, also from plate steel, to visually break up the height of the columns as well as to hide the necessary weld splices of the plates. The column design complements the deep wide-flange beams (white) and the glazing support system, the latter also having been fabricated from relatively standard wide-flange sections.

Custom steel members were required to create the tree-like support system for the atrium of the Atlantic Wharf complex by CBT Architects in Boston, MA, USA (right). Two tapered square-center columns are used to support the skylight system. The branch elements are also custom-fabricated from plate material. The workmanship of the structure is highlighted through applied lighting in this night scene.

The massive truss that restrains the cable-supported glazing system at the Newseum in Washington, DC, USA, designed by Ennead Architects, is created from custom plate steel. The plate provides clean, crisp lines that match the precision of the stainless steel glazing system. The flanges are reinforced with additional plates that have been inset to provide for fillet welding as well as to avoid alignment issues where the web members meet the flange. This detail reduces cost while at the same time improving the appearance of the member. It also allows the webs to work as pairs without requiring being connected to each other to achieve adequate stiffness.

TAPERED TUBES OR CONES

While round tubes only require custom fabrication by a rolling process when they are oversized, all tapered tubes have to be custom-fabricated. Tapered tubes are often used in conjunction with straight tubes to provide a more elegant line or landing at connection points.

Design and detailing with custom steel sections will be discussed and illustrated in more detail in *Chapter 9: Custom Fabrication*.

Barajas Airport Terminal 4 uses custom-fabricated tapered tubes for the large angled members that support the undulating roof beams.

The weld lines in the base detail of the Y-shaped columns at Barajas Airport Terminal 4 reveal its fabrication from parts. The decision was made not to remediate the welds but to allow them to become part of the element, as the scale easily absorbs their impact on the overall appearance.

The tubular legs of the delta frames of QRC West in Toronto, ON, Canada, designed by & Co Architects, are shown here in the fabrication shop of Walters Inc. The tapered portions, fabricated by Kubes Steel, were attached to the round HSS elements. This fabrication team had previous experience in the fabrication of the legs for OCAD. The weld was fully remediated so that it would not be evident after receiving its primer and intumescent coatings. The legs will be filled with concrete, as evidenced by the inlet at the base.

This new bridge near the London Bridge Underground Station in London, England, makes use of custom conical sections within its fully welded truss.

The scale of the members allows for the welds to be neatly done and left unremediated. The separate fabrication of the cones and their connection to the straight HSS members is evident. The geometry of the welded connections has been laid out to provide adequate space for the welding process.

ONTARIO COLLEGE OF ART AND DESIGN | TORONTO, ON, CANADA

The tapered portions at the base of the leg supports on the Addition to the Ontario College of Art and Design (OCAD) in Toronto, ON, Canada, designed by Will Alsop, were fabricated from plate steel via the brake-forming method. The tapered cones were then welded to large tubes. Although simple in appearance, this needs to be carried through as an AESS 4 detail as all of the welds must be filled and ground smooth to make the transition from the cone to the tube appear seamless.

The connection between the tapered section and the tube (bottom left) is fully remediated. The vertical weld seam along the 27m/89ft length of the tube has not been remediated but simply oriented away from the dominant angle of view.

For ease of fabrication, the tapered sections of the legs had their base plates attached (bottom right) prior to having the longer tube attached. A round plate is welded to the top of the cone to provide stiffening at the point of attachment. The light impressions of the brake press can still be seen. These will be remediated prior to applying the primer and intumescent coating.

CONNECTIONS

CHAPTER 7

The transformation of the Las Arenas Bullring Mall in Barcelona, Spain, designed by Rogers Stirk Harbour + Partners, relies on a system of AESS tubes, pin and tension connections to create its architectural statement. Color plays heavily into the expression of the AESS.

The basic issues and solution strategies for connecting steel, including lap joints, butt joints and methods of bolting and welding, are covered in detail in “Chapter 3: Connections and Framing Techniques” *of Understanding Steel Design: An Architectural Design Manual.* The intention of this chapter is to provide a coherent sampling of connection details that can provide the basis of and inspiration for project design and act as visual references. Contemporary connection design has borrowed significantly from the type of detailing characteristic of the High Tech movement. More information on these precedents can be found in “Chapter 5: AESS: Its History and Development” in *Understanding Steel Design: An Architectural Design Manual.*

DEFINITIONS

Member refers to the discrete sections of steel, such as wide-flange (Universal) sections, hollow structural sections (HSS), angles, channels, rods or cables.

Element references the larger agglomerated pieces of a building. This includes trusses, beams and columns as they extend from one external connection point to the other. A small or uncomplicated element may be constituted simply by one steel member. In many AESS projects the additional complexity will require the assemblage of larger elements from a number of members.

Connections are of three basic types by virtue of their location and purpose:
Internal connections are those by which the members are joined to create a larger element. These are most normally the result of shop fabrication.

External connections connect elements to each other. These are most often completed on site. This includes, for example, the connection of a truss to its supporting column or a beam to a truss.

Splices are to be found when elements are too large to ship in one piece. These are often completed on site, either on the ground prior to lifting or in the air as erection proceeds.

Connections located within an AESS element tend to be done in ways that suppress the evidence of the connection. Connections between AESS elements will choose the level and nature of the expression of the connection. Where concealed structural steel design tends to select the most structurally effective and economical connection type, AESS has aesthetically dependent variables with which to contend.

Splices have different characteristics in an AESS project. These are special connections that discreetly happen within sections of elements where the act of connecting is intended to be hidden. These connections often arise as a result of the inability to transport oversized members, hence requiring the element to be fabricated in smaller sections that are aggregated on site. They will be addressed separately in *Chapter 8: Specialized Connections*.

Standard structural shapes such as wide-flange (Universal) sections, tubes, angles, channels and plates base their AESS connection details on variations of standard structural detailing. By contrast, extremely slender members such as the cables and rods that are used in tensile structures require specialized connections. These will be addressed in the last part of this chapter.

CONNECTION DESIGN CHECKLIST

The issue of connection design is highly variable as a function of the wider range of choices inherent to the architectural use of structural steel. Initial choices impact subsequent choices, creating a domino effect. Aesthetic drivers in AESS influence these choices.

In the relationship between the AESS category and the specific connection detailing, it is an open question which comes first, the selection of the AESS category or the desired detail/aesthetic. In some cases the choice of the AESS category as a function of viewing distance, use of the building and budget will lead to the selection of appropriate connection details. In other instances the desire to use certain connection types will infer a category preference. In this chapter details will be discussed in relation to their appropriate AESS category as well as to their fabrication and erection requirements.

Particular AESS connections demand a category of AESS with regard to the fabrication requirements for achieving a certain appearance (independent of distance to view). The taper on this strut must be specified AESS 3, because the connection between the parabolic plate that creates the taper and the tube must have the welds ground and filled. The distance to view would have to be improbably large to allow for an unremediated weld condition.

Analysis of the following questions can direct and inform the connection design on a project-by-project basis:

1. *What type of members are to be used?*
   Connection types are impacted by the choice of structural section. Hot-rolled steel shapes (such as W, Universal, C and L-shapes), those that use hollow structural sections (HSS) and those making significant use of plate material have differing requirements.
2. *Does the design make use of specialty items?*
   Rods and cables as in tension systems require the use of special means to create connections, such as clevises, turnbuckles and swages. Smaller castings for tensile connections or larger expendable mold castings require different detailing and incur additional expenses. Curved members require extra expense in fabrication to bend or roll the steel.
3. *Is there an aesthetic desire for bolting or welding?*
   While bolting and welding can create similar structural capabilities in connections, their impact on the nature of fabrication and erection is extremely different. Sometimes bolting and welding are desired for their respective aesthetic qualities. This raises different issues beyond structural effectiveness.
4. *What is the budget regarding the AESS category/categories targeted?*
   Budget will have obvious impacts on the nature of the fabrication work. It limits or allows remedial work such as filling and grinding, which in turn has repercussions on the connection design. Budget can direct a decision to choose an AESS category, whose associated characteristics can in turn lead to budget decisions. If a project can be subdivided into different categories of AESS, then the more visible areas can be more expensively detailed.
5. *Globally speaking, where is the project?*
   Fabrication preferences, methods and expertise vary substantially around the globe. The labor rates in different markets will directly bear on the budget. Highly articulated work may be beyond the capabilities of the work force at hand. Certain countries make widespread use of site welding (e.g., China) while others strive to avoid it (e.g., Australia). Knowing the local market is critical to detailing a project within the local expertise. Understanding these limitations will help to avoid disappointment and conflict. Shipping distances can play into location and budget.

The Rose Center for Earth and Space in New York City, NY, USA, designed by Ennead Architects, uses a highly complex tubular truss structure to support its glazed envelope. A typical connection was designed to terminate the round HSS members that were connected on site. The connection is the same in a three-dimensional formation (left) or a planar one (right). The hinge connection employs simple bolting to a plate tab shop-welded to the primary truss sections. The closure of the end of the tube with a circular plate requires an AESS 3 category as the weld between the plate and tube needs to be ground smooth. By contrast, the welded connection at the internal right angles of the tapered plates did not need remediation because it is out of direct view by virtue of the geometry of the connection. Where bolts are used as fasteners, care has been taken to keep the heads located consistently.

A clear understanding of the guiding questions should inform the design of the internal, external and splice connections on a project. The selection of the connection type depends on the structural requirements of the assemblies, the shapes and types of steel members that are to be connected as well as the desired aesthetic. Preferences may not be clearly evident from the outset and it may be necessary as well as desirable to use different types in a project. Considerations of the desired aesthetic will lead to choices of welding over bolting, or vice versa. Generally speaking, welded connections result in a more smooth and continuous appearance of the structure, while bolted connections provide a more technical appearance with a clearer indication of the method of assembly and definition of the discrete elements that comprise the structure. A consistent approach to the connection detailing within a project might create "detail families" that share a common language even if the nature, location and geometry of the connections differ.

SHOP VERSUS SITE FABRICATION

It is mentioned repeatedly that to weld in the shop is easier than on the site. This is not to infer that site welding is not done. It is simply to raise the awareness that site welding will lead to cost increases and, as a function of the complexity of the connections, to longer erection times.

This connection at the base support of a truss element at the Jay Pritzker Pavilion in the Millennium Park in Chicago, IL, USA, designed by Frank Gehry, shows the use of welding in a connection. The shop and site welding appear differently. It is a challenge for the welder to complete the vertical weld where the truss is attached to the plate that has been set in the concrete wall. There are also logistic and access challenges in the round HSS member coming into the top of the connection when attached on site.

Therefore, as a general rule it is better to maximize the number of connections, particularly to maximize the welded connections, that can be done in the shop over those that must be done on site. There is more environmental control in the shop. Jigs can be set up for repetitive assemblies to ensure consistency in appearance and finish. Equipment is on hand to turn and lift the steel, providing easier access for welding. For the same reasons, the application of primer and even finish painting is also more efficiently done in the shop.

In the end, the maximum member size that can be transported to the site often determines the scope of shop-fabricated connections and the number and type that must be done on site. Although it is possible to transport oversized pieces to the site with an escort, this does increase the cost of the project. Likewise, when the site is constricted and the staging area either small or non-existent, pre-assembling of pieces on site might not be possible and connections will need to be completed "in the air" during erection. Most urban sites require "just-in-time" delivery of steel pieces and carefully planned erection.

Size limitations imposed by transportation create the third category of connection, the splice within the length of a (long) element. Typical are large-spanning truss elements where the admissible transport length falls well short of the span length. Often an obvious internal splice connection is not desired as it would ruin the lines of the form. Splices can be designed to be fully exposed, discreet/hidden or fully concealed. These will be discussed more fully in *Chapter 8: Specialized Connections.*

This 76m/250ft-long truss uses bolted splice plates in the top and bottom chords to accommodate shipping constraints. The diagonal ladder-like web member was bolted into place at the time of splicing. The sections were delivered to the site and laid flat to permit easy access to complete the full assembly of the length of the truss prior to lifting. The balance of the connections were shop-welded. The project makes predominant use of standard steel sections; it would fall into the AESS 2 category.

CONNECTION MOCK-UPS AND VISUAL SAMPLES

Referring to *Chapter 3: Characteristics* (see page 46), the importance of visual samples plays heavily into the ultimate decisions on the nature of the fabrication of the connections. Ideally a designer or client would desire to see and feel specialty connections before they commit to their mass fabrication. This is not always possible or practical due to issues of timing and cost. Depending on the scope of the project and the quantity of AESS to be incorporated, the contract for the steel fabrication and erection is often awarded after the project has been designed and the architectural contract documents completed for the bidding or tender process. This can mean that the fabricator has not had the opportunity to provide helpful input into the design of the AESS system and its connections. If the contract is not awarded to a fabricator with extensive expertise in AESS, this can cause problems with mismatched expectations. Although the use of the AESS Category System should preclude most issues, it cannot solve all questions arising during the design detailing of the connections.

For large AESS projects it is not uncommon to pay for fabricator consultation during the design detailing and contract document phase of a project. The consulted fabricator will then usually bid on the work, with the understanding that award is not guaranteed.

The fabrication of large specialty items is expensive and time-consuming. The preparation and assessment of physical mock-ups can create delays. Viewing distance needs also to be taken into account; normally the sample is examined at close range when in fact the in-situ connection may be many meters out of range of view and touch. The AESS category must also be kept in mind when viewing physical samples.

It may be possible to verify most of the appearance issues through the use of 3D drawings, saving time and money by relying on a combination of the drawings that are produced by the fabricator's detailing software and ones that are produced with 3D modeling software. If 3D or other sorts of digital models are used as the basis of agreement for details, it is important to discuss the finer aspects of welding, bolting and finishing, as these are likely not represented fully in the digital model. A combination of smaller physical mock-ups of aspects of detailing and finish might be used to accompany digital representations to achieve a good level of communication about the expectations of the connection details.

BOLTED CONNECTIONS

There are two common types of bolts used in bolted connections, Hex Head bolts and Tension Control (TC) bolts. Rivets are no longer used in contemporary construction. Where historic projects that use rivets are being renovated or refurbished, TC bolts can be used as their round heads mimic the appearance of the rivet head.

Various types of steel can be used as bolt material:

- High-strength steel is most commonly used for its superior load-carrying capacity.
- Stainless steel is used in applications where appearance is important and weather is an issue.
- Galvanized bolts are used in conjunction with galvanized steel projects.

The majority of bolted AESS connections are designed as hinge connections because they are intended to transfer shear forces and are not required to resist bending moments. Hinge and pin connections are differentiated by the number of bolts used, which in turn impacts the ability of the connection to rotate. A pin connection needs to be able to rotate, so it comprises a single bolt or cotter pin that has to achieve the shear transfer. A hinge connection has a greater number of bolts in order to resist the shear forces; it is stiffer and unable to rotate, but it is still not intended to resist any bending moment.

Bolted connections are normally chosen for their more rugged aesthetic and more industrial look or to deal with erection issues and constraints. They are often combined with W (Universal), C or L-shaped profiles. Often the detailing of bolted AESS connections is very similar to the same connections in standard structural steel, with the exception of an enhanced concern for geometry and tighter tolerances.

Even a fairly standard bolted connection can be elevated to AESS quality by careful design and tighter tolerances. The Ronald Reagan National Airport in Washington, DC, USA, designed by Pelli Clarke Pelli Architects, uses wide-flange sections with bolted connections. Care is taken to keep the tolerances tight and to ensure that the bolts are carefully spaced and the heads located consistently. Where the railing penetrates the web of the beam, the bolt spacing has been adjusted to suit. The tighter tolerances make this type of detailing more difficult to erect and require characteristics of AESS 3 category.

It is part of the responsibility of architectural design to specify the types of bolts and assure the consistency of the side on which the bolt head is to be found. As the structural requirements of the bolted connection dictate how far a bolt needs to be tightened, it is not reasonable to expect the rotation of all heads to align.

Although some types of bolted connections allow for adjustment on site, it is not always desired to have slippage. In complex geometries the steel sections have to be designed to fit (hence half standard tolerance as a minimum requirement) as slight amounts of misalignment can easily accumulate and result in an inability to fit highly complex steel members.

In this detail of the Dalhousie School of Architecture in Halifax, NS, Canada, the end plates welded to the adjoining wide-flange (Universal) sections have been trimmed more tightly to the face of the connection. The bolting is predominantly located on the side extensions of the plate.
The connection has been dimensioned to permit tightening of the bolts in the gap between the plate and member. This can be achieved with an AESS 2 specification even though this detail is located close to view.

In this connection at Heathrow Airport Terminal 5 in London, England, designed by Rogers Stirk Harbour + Partners, stainless steel cotter pins that are more commonly used in pin connections have been used instead of Hex Head or TC bolts. This choice of materials creates a different aesthetic for the connection. The alternate connectors combined with custom-cut grey plates elevate the connection to an AESS 3 category.

Bolted connections can also be used with hollow structural members. Typically a plate is welded to the HSS member to facilitate bolting. The plate can be welded to the end of the hollow member, allowing bolting through the projected plate edge, or this can be accomplished by inserting a plate in line with the member.

Two types of bolted connections have been employed in the connections of the square HSS members for the Canadian War Museum in Ottawa, ON, Canada, designed by Raymond Moriyama Architect. One type uses a set of overlapping plates (at the X intersection) that are inset into a slice in the tube; the other uses a simple connection where the plates are welded to the ends of the HSS members and then bolted. The aesthetic of the space, mimicking a twisted war-torn landscape, inspired these connections. The tight tolerances and the desire to control the orientation of the weld seam on the HSS correspond with the AESS 3 category. The use of end plates on the HSS allows for a degree of expression of the connection that makes the structure feel rougher, in spite of the glossy appearance of the painted finish. The use of exposed steel decking assists in providing an industrial texture to the space. Fabrication and erection by Walters Inc.

Simple end plate connectors have been used to attach the long triangular truss sections of this pedestrian bridge in Auckland, New Zealand. These contrast with the internal truss connections, which have been welded. This type of connection simplifies erection. This end detail could be classed at AESS 2.

Bolted connections with simple lap joints constitute the internal connections in this pedestrian bridge truss. The attaching plates have been cut into the round HSS members. The ends of the tubes have been cut on an angle and completed by welding a semi-circular plate to prevent entry of moisture. The welds have been left exposed and although the geometry and alignment are controlled, the plates are allowed to protrude, thereby providing a substantial cost saving. This type of detail corresponds to the AESS 2 category.

WELDED CONNECTIONS

Welded connections are used in many AESS structures, often in conjunction with hollow structural sections but also with more standard structural shapes. Welding gives a clean, uncluttered appearance. Specifications regarding grinding of the welds constitute a critical difference between AESS 2 and AESS 3 categories; in AESS 1 and AESS 2, welds will be left unremediated. At the AESS 3 level weld remediation, typically achieved by grinding, is considered optional but not always necessary – weld remediation can add significantly to the cost of the project, particularly if welding and weld remediation are to be done on site. At AESS 4 level the requirement for blending and contouring of connections typically leads to weld remediation and the majority of connections, both within and between members and elements, are welded.

Welded connections present specific problems for the fabricator as a function of the geometry of the connection and the choice of member. For complex geometries to be more affordable and for better quality and alignment, it is necessary to maximize the amount of work that can be done in the fabricator's plant so that proper jigs, lifting and clamping devices can be used to manipulate the materials. Where highly articulated assemblies must be broken into smaller elements due to transportation and lifting limitations, the details of the more significant site connections need to be discussed with the fabricator, if a totally welded appearance of the structure is the desired result. It is possible to create site connections that convey the appearance of being welded but are discreetly bolted and where the final connection is concealed with cover plates. These are discussed in detail on pages 140/141.

For the decision on the level of finish of a welded connection it is extremely important to take into account the viewing distance and the intended AESS category with its associated characteristics. One of the major reasons for cost overruns in design and construction with AESS has been the tendency of welded connections to be overworked. Welds are often unnecessarily ground, filled or smoothed out. Overgrinding can even diminish the structural strength of a weld. Except in the instance of structural necessity, or for seal welding to prevent moisture entry, there is often not even the necessity for welding to be continuous. As a rule, grinding should be reserved for Custom or high-end AESS 4 welded connections.

If deciding to leave a welded connection unremediated, there needs to be good communication with the fabricator as to the expectations of the designer and client, allowing that:

- Much welding is done by ironworkers, so it is craft-based; therefore there will be minor inconsistencies from weld to weld as a function of who was doing the work, the ironworker's experience level and the ease of access to do the welding.
- It is likely that shop welding will be superior to some site welding as a result of access issues.
- The design of the connection and choice of member sizes must allow for the physical space of the weld.

Not all welded connections present the same challenges in detailing.

- The most basic weld between members can be done using a straight cut on the end of one member and connect it to the flat face of a joining member by simple fillet welds.
- Connecting tubular members to other tubular members requires more precision, as these often include odd geometries and welds that are not straight.
- Butt welds are used to create joints between members or surfaces that are intended to remain in line. The choice here is to conceal the splice (see page 56) or allow it to be exposed (see *Chapter 8: Specialized Connections*, page 140 for more detailed information).
- Interior corners can be connected using a fillet weld. The remediation of this weld is more difficult when the angles between the members are acute. However, remediation is often unnecessary as the placement of the weld tends to make it less obvious.
- Exterior corners present a wide range of detailing choices. If the corner needs to look sharp, welds must be done and ground smooth, removing all evidence of the weld and leaving the impression that the steel is a single piece. If the welds on this type of sharp corner are left unremediated, a rougher look will result. If a corner is created by insetting one plate from the other, fillet welds can be used that can be left unremediated.

This straight-cut tube (top left) is welded to the flat face of another member. When creating tube-to-tube welded connections, it must be taken into account that rectangular or square HSS do not have perfectly square corners. Their corners are rounded as a function of the thickness of the steel plate used to fabricate the section. Enough space must be left for the weld. If the adjoining round tubes had the same exterior dimension as the square HSS, there would be considerable bleed of the weld over the edge. This connection would be considered as an AESS 2 example.

Another example of a straight-cut tube welded to the flat face of another member (bottom left), also of AESS 2 level. The adjoining HSS members of this truss have been arranged to allow adequate space for welding. The web members are slightly smaller than the chords so that the fillet welds fit without bleeding over the shoulder of the rounded edge of the HSS.

A butt weld connection between joining HSS members (top). The aim was to join identically sized intersecting square HSS members with a flush condition. The joint needed to be prepared and the weld to be selected to allow for this level of grinding. This connection would qualify as an AESS 3 example.

Tube-to-tube connections become more complicated when it is not possible to use a straight-angled cut to create the joint. Even the invention of CNC and other precision cutting equipment has not yet obviated careful craft in this process.

The cut lines are marked on this tube in preparation for cutting. It is at this point that the fabricator must account for the control of the position of the natural weld seam on the member.

The tubes are tack-welded to align with further fabrication marks that have been drawn onto the chord of the truss. The web members are slightly smaller than the bottom member in order to allow for neater welding.

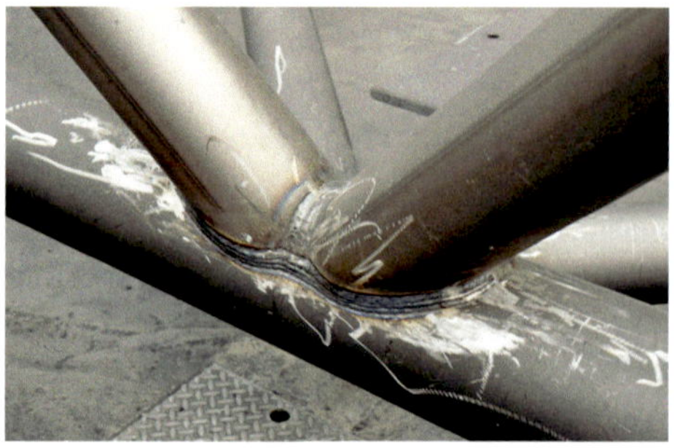

Having the tack welds in place makes the final welding easier to accomplish. There is a significant amount of weld on this connection. In the example of this bridge, there will be multiple coatings and so the location of the weld seam has not been controlled, as the thickness of the coating system as well as distance to view do not warrant such extra attention.

Great care is required when considering remediating welds that are located in tube-to-tube connections, particularly of round members. It is extremely difficult to grind evenly in valleys. Filling with body filler and subsequent sanding, refilling and sanding will be required. If this treatment is to be performed on a high number of internal connections, this can add greatly to the cost. Always consider viewing distance and that much of the weld detail will not be perceived by the normal occupant.

Roy Thomson Hall in Toronto, ON, Canada, designed by Arthur Erickson, completed in 1982, allowed the welded connections to remain unremediated. This project is an example of very early use of AESS that predates any computer-assisted cutting of the members. The overall conical shape of the tubular skylight support system was geometrically consistent, resulting in fewer variations in the design and fabrication of the X-shaped connections. The cuts to the round tubes are straight, resulting in fairly simple connection welding.

A close view of the connection shows the careful use of welding that is quite uniform from connection to connection. This decision afforded considerable savings. This could still be classed as an AESS 3 type connection – but one that did not ask for weld remediation.

The use of curved steel required a different approach to creating the connections between the round tubular members of the pedestrian bridge at the Museum of Flight in Seattle, WA, USA, designed by SRG Partnership Inc. and Magnusson Klemencic Associates. The round shape of the hoops remains intact in the leftward leaning direction. The hoops angled to the right are fabricated from segments of steel tube that have been cut to fit and create the lattice appearance. Cutting the segments from continuous curved pieces of steel is required to assure a good fit and consistent geometry.

This detail of the intersection of two round HSS members shows that the welds have been neatly done and left unremediated. The cut to the joining members is more difficult as it is curved. The interior angles between the members are wide, making welding easier to access. This would qualify as an AESS 3 type connection.

The curved tubular trusses that define the sky level restaurant of the 441.8m/1,449ft tall KK100 Tower in Shenzhen, China, designed by TFP Architects, are designed to AESS 4 level. The exquisite and expensive nature of this space warranted this level of remediation.

A close view of the internal member-to-member connections of the tubular truss the of KK100 Tower highlights the level of remediation achieved. Substantial grinding and filling was necessary to completely mask any evidence of the welded connections.

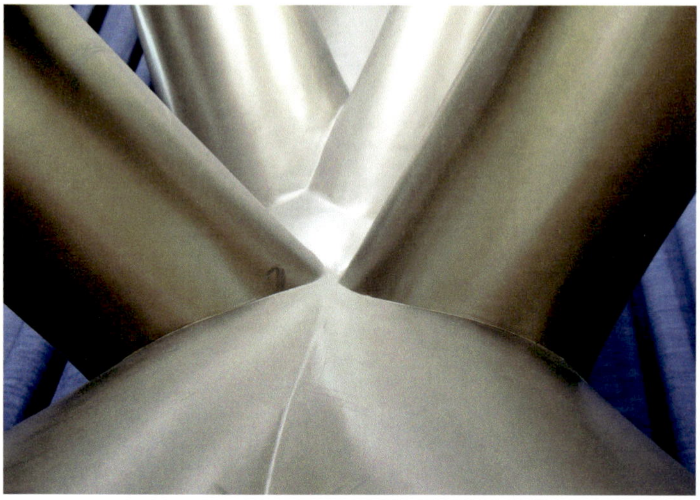

The HSS exoskeleton (top left) of the Torre Diagonal 00 Telefónica in Barcelona, Spain, designed by Estudi Massip-Bosch Arquitectes, required weld remediation to allow the steel frame elements to achieve a clean sense of continuity.

The angled connections between the large rectangular HSS tubes (top right) at the Torre Diagonal 00 Telefónica have been fully remediated to remove any trace of the connection. Remediating the face of the connection would have been easier than finishing the acute interior angle. However, the weld seams on the tubes have not been remediated, but as the seams are consistently facing the inner side of the member they are not very apparent. This could be classed as AESS 3.

The curved box sections for the Arganzuela Bridge in Madrid, Spain, designed by Dominique Perrault (left), have been fabricated from bent plate. The plates have been joined along their length using butt welds. The scale of the bridge and dappled lighting allow the unremediated welded connections to recede from view, in spite of being at close viewing range. Given the amount of site welding required to erect the shop-fabricated elements of the bridge, the decision was taken to exclude weld remediation and leave the exposed welds as part of the aesthetic of the bridge. This level of custom fabrication is a direct result of geometries that preclude using off-the-shelf steel members. The major design decision rests in the fabrication of the "corners" of the curved boxes. The desire was to keep them as square as possible.

Welds used between plates to create larger aggregated elements fall into the Custom AESS category and will be examined in more detail in *Chapter 9: Custom Fabrication*.

For the custom-fabricated truss in the Newseum in Washington, DC, USA, designed by Ennead Architects, the plate that joins the two sides was recessed rather than keeping it flush and grinding away the weld. This creates a more simple detail to fabricate and also adds to its aesthetic expression.

PIN CONNECTIONS

Pin connections are frequently used as element-to-element connections as they can simplify erection on site. Pins are technically considered to be bolted, although AESS applications normally use specialty connectors, pins in this case, that are more aesthetically refined and very lightweight in appearance. Pin connections are structurally designed to transfer shear forces and are not designed to resist bending moments. Sometimes they are designed to accommodate a certain degree of rotation, be it that the joint moves like an actual door hinge, or that the joint can accept some degree of rotation as a function of the erection process.

Pin connections can either be custom-fabricated from plate when associated with the connection of large structural members, or simply use clevis connectors. They can be used in conjunction with extremely thin members such as rods and cables, or with larger structural types.

This detail at the Las Arenas Bullring Mall shows a combined use of clevises and more standard plate-fabricated pin connections in the tension system. The structural hierarchy is made evident in the thicknesses of the attaching plates and the nature of the welds used to join the plates to the central round post. The clevis connections are in tension, while the plate connectors are providing side-to-side bracing.

Specialized connections are often required when attaching tensile members to an exposed steel system. These can be custom-designed and fabricated to match the architectural language of the project, or they can be selected from off-the-shelf sources. There are four primary areas of consideration to be aware of when designing tension connectors:

- This is normally a pin connection, so transfers no moment.
- The rods or cables are normally quite small in diameter and must be securely fastened to resist pull-out due to tension.
- The type of connector will vary as a function of the use of a cable or rod: rods can be machined to have threads where cables cannot.
- These systems are normally installed "loose" and must be able to be tightened in order to secure, tension and plumb the systems.

Many tension connectors are made from cast steel and mass-prefabricated as the demand is quite high. These less expensive products can easily be incorporated into the design. In many cases, and with very large or expansive structures, custom-fabricated connectors will be required that provide a better solution in terms of aesthetics and functionality.

While it is straightforward enough to specify a simple clevis attachment at the end of a rod or cable to provide the load transfer, detailing becomes more challenging when the load transfer point must accept multiple incoming members. The idea behind tension structures is to keep a lighter than normal appearance, so the connection details must reflect this basic idea. The erection process needs to accommodate tightening and plumbing. Depending on the overall design of the system and the connection devices chosen, this can be done at the termination points or along the length of the rod or cable by using a turnbuckle.

By the nature of the work involved in the fabrication and erection of tensile systems, these are likely to be associated with AESS 3 and AESS 4 projects.

Pin connections are useful to resolve odd geometries as well as the connection of a variety of incoming members. This connection is part of a high-level roof support system. The evidence of fabrication is apparent as many of the welds have not been remediated, which is quite acceptable given the distance to view. Water is prevented from entering the hollow members. The square members must be custom-fabricated to achieve the sharp square corners.

A custom pin connection requires the attachment of one or more plates to the end of the member to create what is essentially a clevis-like arrangement. The number and thickness of the plates will vary to suit the amount of load that must be resisted. This is also true for the sizing of the pin or bolt, whose cross-sectional area must be sufficient to resist the shear forces. Much in the creation of this detail depends on how the plate is attached or inserted into the primary member.

This pin connection uses a pair of plates set into a tubular member. The taper of the tube creates a parabolic shape that will be infilled with a plate cut to match. These pieces will be welded together and the weld fully remediated to make for a very clean appearance. An extra thickness is added to the plate to account for the load transfer. This also adds an interesting detail to the completed connection.

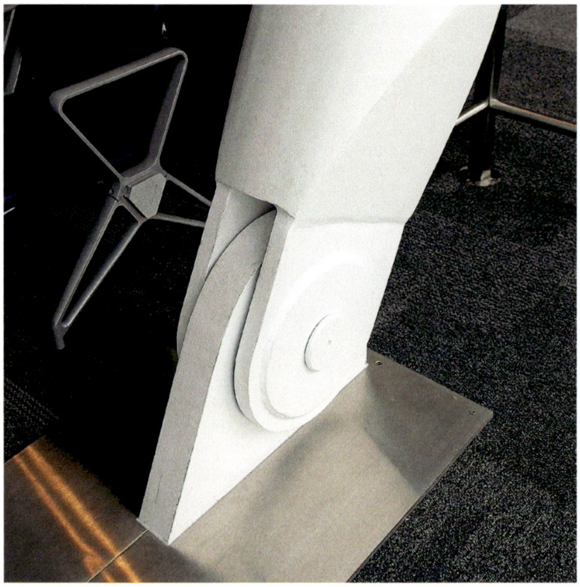

The finished element at Pearson International Airport in Toronto, ON, Canada. The large flat round "head" at the center of the connection is a special type of cotter pin. It can vary in diameter to suit the load transfer. The cap may have one, two or three holes (more holes for the larger cap). These connections are used in conjunction with a special tool to tighten the cap into the end of the pin. The nature of the finishing of this detail aligns it with AESS 3.

This version of a pin connection would be much easier to fabricate and could be accomplished at an AESS 2 category level if the end of the tube were left open. The tack welds holding the plates in place would be completed using a fillet weld and left unremediated given the extension of the plate beyond the edge of the tube.

In this double pin base connection the plates have been designed to extend beyond the tubular columns. This allows for simple fillet welding and minimal remediation. The ends of the tubular columns are filled in with a semicircular plate whose welded connections have been ground smooth. This would be classed as an AESS 3 detail.

When clevis-type pin connections are used to resolve large quantities of connections in areas of limited attachment, care must be taken to ensure that the plate is sized to allow enough space between the holes. The plate thickness may need to be increased to accommodate load transfer and eliminate pull through shear forces.

The base of the tensile roof structure at the Sony Center in Berlin, Germany, designed by Helmut Jahn, is an excellent example of careful planning of the plate-to-post geometries to allow for welded attachment. The plate thickness at the point of clevis attachment has been increased while the transition between the thinner plate, required to accommodate the attachment to the post, and the thicker plate has been beveled to appear more graceful.

The mast attachment (right) for the Jubilee Bridge in London, England, designed by Lifschutz Davidson Architects, has a sawtooth profile on the lower edge of the plate to accommodate the numerous clevis attachments.

The plates on another bridge mast (left) are articulated to permit the distribution of the clevis attachments, ensuring that their load paths are properly directed and that there is an adequate amount of steel between the holes.

BARAJAS AIRPORT TERMINAL 4 | MADRID, SPAIN

Barajas Airport Terminal 4 in Madrid, Spain, designed by Rogers Stirk Harbour + Partners, employs a family of pin connection details that are each suited to location and distance to view. The project relies on a high level of prefabrication and a preference for making on-site connections via bolting.

This pin connection (left) is very simply finished, appropriate to its high location at the ceiling level of the departures hall. The rounded plates have received a simple bevel cut so that the surfaces align on the exterior when the plates are inserted into the cuts in the tube. The welds have been neatly done and not remediated. The upper end has been left open as it cannot be seen from below and the interior location of the connection protects it from water penetration. This would be classed at an AESS 2 level.

The pin connection at the top of the arch (right) is quite simple and fabricated to AESS 2 standards. The connecting elements are attached to the curved beam ends via bolting. The detail is situated well above viewing distance.

The detailing is consistent from interior to exterior. Color was additionally used to assist with comprehending the length of the terminal building.

Custom connections were used to secure the horizontal ties that stabilize the two arms of the Y-shaped support system (left).

Pin connections were also used at the base of the Y-shaped supports (right). These allowed easier prefabrication and erection as they permit natural rotation during the alignment process.

The higher degree of load transfer required the use of additional plate elements in this pin connection at the University of Guelph Science Building in Guelph, ON, Canada, designed by Young + Wright Architects with Walters Inc. The high ceiling level location allowed for the use of a simple high-strength bolt to create the pin. The TC bolts that connect the plate to the beams above help to create a less technical appearance while facilitating erection.

The end condition of this member indicates the custom nature of its fabrication. A taper always requires innovative fabrication. In this instance the entire tapered end, including the hinge, has been machined from a solid block of steel. The piece was welded to the tube. The slight step detail allows for ease of connection as there will be very small deviations in "true round" between the pipe and the machined element. This would qualify as AESS 4.

This connection detail has been very simply refined by rounding the corners of the plate tabs. The round plates that terminate and seal the ends of the round HSS members are sized to permit a connection via fillet welds. The connecting tabs are fillet-welded to the round plates. These very simple pin connections would fall into an AESS 2 category.

The structural supports for the Dubai Metro Stations, UAE, designed by Aedas Architects, use a large pin connection to join the ribs to the base support. The slightly V-shaped voids between the members indicate that some alignment was made possible during erection. The rib features crisp, square, fully remediated welded corners and the use of custom plate box sections. Insetting the plates at the sides of the trapezoidal base has allowed a fillet weld that has not required remediation. Clevis connections are used to terminate the ends of the cross-bracing members.

There are numerous interesting possibilities when designing with tension connectors in force-differentiated AESS structures. It is always advisable to discuss the options among the team members as decisions will need to be taken if using custom-fabricated connections in lieu of more standard clevis arrangements. In all instances, the plate to which the clevis connects will require careful consideration to accommodate geometries and loads.

The exterior bracing system at Neo Bankside Housing in London, England, designed by Rogers Stirk Harbour + Partners, employs a family of connection details that use plate-fabricated pin connections as well as clevises.

At the upper end of the elliptical tube support (top right), the plate has been cut into the narrow end of the tube and the welds have been remediated. Unavoidably, a portion has been flattened. The distance to view makes this detail unapparent. This would be classed as AESS 3.

This pin connection at the base of the exterior steel frame (bottom right) has a high level of finish, with regard to its proximity to view and the exterior condition. The connecting plate has been inserted inside the elliptical tube and the welds have been fully remediated.

BAR

SPECIALIZED CONNECTIONS

CHAPTER

8

Special castings form the pin connections for the tapered columns at the expansion to Zurich Airport, designed by Grimshaw Architects.

Beyond the adaptation of basic bolted and welded connections to suit AESS applications, there are other types of connections that respond to sophisticated expectations on the act of connecting, however without requiring completely customized fabrication as will be addressed in *Chapter 9: Custom Fabrication*. These special connection types are often used in conjunction with the range of bolted and welded connections described in the previous chapter. They address design and erection issues that include splicing and complex geometries arising from the convergence of multiple members.

MAKING CONNECTIONS LESS VISIBLE

The majority of connections used in AESS structures tend to enhance, or make a design detail out of the act of connecting the steel. In certain instances, the design intention will be to hide connections or make them less visible. This is typically the case when splicing steel elements and it can also apply when trying to make connections between members appear more trim.

The limitations on the transportation size of elements will often require elements to be shipped as a number of smaller pieces, so that site connections have to be carried out. In AESS projects, particularly those designed with an all-welded aesthetic, this can present erection issues, as on-site welding can be problematic and bolted site connections are liable to disrupt the visual lines of the elements.

Hidden or discreet connections offer an alternate to fully welded connections, in particular when it comes to splices within structural members (typically HSS) that are too large to be transported in one piece. Hidden bolted connections are often done to splice the segments of long-span trusses on site prior to lifting, in order to allow the lift of the fully assembled element and avoid the need for temporary support or stabilization systems. In a hidden connection, the bolts are concealed by a cover plate. In a discreet connection, the bolts are apparent but the connection is designed to make it less obtrusive.

Hidden and discreet connections are able to save significant time and money in fabrication and erection, while achieving an aesthetic that is similar to a welded connection.

HIDDEN CONNECTIONS

Bolted connections can be hidden beneath specially designed cover plates that are shaped to match the form of the primary attaching members. The splice plates are attached to each end of the joining members. They overlap and are bolted together in order to form the structural splice. The physical requirements for the splice plates and bolts are simply to fit neatly within the cover plates. At interior locations the cover plate can be tack-welded into place, filled with body filler and sanded prior to applying coatings. On the exterior of a building great care must be taken to ensure that the cover plate is sealed against the entry of moisture. This type of hidden connection would be classed as AESS 2.

The long trusses of the International Terminal at Calgary International Airport in Calgary, AB, Canada, designed by DIALOG with fabrication and erection by Supermétal, arrived on the site in three sections. While the top chords were made structurally continuous through the use of pin connections, the bottom chord was spliced using a hidden connection. As the bottom chord was very visible in the finished state, the hidden connection was important for the appearance of continuity.

Shop test for the alignment and fitting of a hidden connection for the International Terminal at Calgary Airport. The bolted splice plates are designed to sit completely within the connecting tubes to facilitate the addition of cover plates.

The AESS 2 truss at the International Terminal of Calgary International Airport uses a hidden connection to splice the bottom chord. This is very time-effective and for the given distance of view quite an acceptable solution. Once the finished ceiling is in place, the pin connections of the top chords are in limited view whereas the bottom chord is in a dominant view position.

Detail (top) of the bolted connection inside the cover plates.

The ANZ Stadium at the Sydney Olympic Games site in Sydney, Australia, designed by Populous and Bligh Lobb Sports Architecture, uses hidden connections to join the sections of the truss that ring the circumference of the stadium.

The cover plates over the hidden connections (top) blend in well with the tubular members. The main "branches" to which the truss sections connect have been welded, with the welds left unremediated as an acknowledgement of the distance to view and the scale of the truss.

DISCREET CONNECTIONS

In discreet connections between tubular members, the bolts are left visible but the connection is designed in such a way as to reduce the visual impact and retain the visual lines of the joining members. Two basic methods are commonly used. In the first method, splice plates are inserted into the ends of the joining tubes and the overlap is bolted. This places the bolts and splice plates more or less within the dimensions of the joining members. Discreet connections are normally fabricated to AESS 3 standards as a function of the amount of remediation/grinding required when the plates are attached to the tubular steel.

A discreet bolted connection is used in this canopy at Baltimore Washington International Thurgood Marshall Airport in Baltimore, MD, USA, designed by SOM with URS Corporation and AECOM. The splice plates maintain the visual lines of the tubular truss members and the bolts are set into the gap. The overall effect could have been improved by applying a characteristic of AESS 1 – locating the bolt heads on the visible side of the connection and the nuts on the reverse.

The truss splice of the canopy at the BWI Airport is in keeping with the visual intentions of the welded truss, without incurring the extra expense for welded connections.

The connections used between these square HSS elements in the Canadian War Museum in Ottawa, ON, Canada, designed by Raymond Moriyama Architect, showcase a combination of a standard use of matched end plates and a discreet connection where plates have been cut into the tubes and bolted using a simple overlap of the joining plates. The ends of the HSS members have been capped using a plate that has been welded and remediated to provide a smooth look in the AESS 3 range.

The second way of joining HSS material consists of completing the ends with a pair of plates whose dimensions match those of the tube. The plates are sectioned and fitted with bolt holes. Care must be taken when cutting the end plate and welding it to the tube or custom section to properly size, machine and remediate the welds so that the plate appears to be an integral part of the end of the tube. This level of fabrication puts the connection at a minimum of level AESS 3. Each joining member may be fitted with this connection; if there is only one of these connecting elements it is bolted to a simple plate. Generally this connection will require access to both sides for bolt tightening.

The support column is attached to the branches of this "tree" on a building designed by Rogers Stirk Harbour + Partners at Potsdamer Platz in Berlin, Germany. A discreet bolted connection has been used.

This type of discreet connection must allow for bolting, with consideration of the size of wrench or other equipment as well as any maneuvering or turning that is required to tighten the bolts. Plates subdivide the column diameter into segments to complete the load transfer, with one segment per bolt – effectively creating a series of chambers. The Hex Heads of the bolts can be seen from the bottom of the connection, which is of course the dominant view angle.

A discreet connection is used on this rectangular box beam-to-plate connection at the Art Institute of Chicago in Chicago, IL, USA, designed by Renzo Piano Building Workshop. This sort of connection can be adapted to various locations and member shapes. The niches for the bolts are quite small, making bolt tightening more difficult.

AESS is used in the wind tower in Masdar City in Abu Dhabi, UAE, designed by Foster + Partners. AESS 3 is required to ensure a seamless transition between the end plate and the tubular member including grinding of the weld. It is critical to ensure that the small plates that divide the connection into chambers align between connecting members. The chamber size must also allow for welding that is not remediated as it will be hidden.

The base plate has been set flush into the pavement and the bolts have been recessed. The structure uses round HSS members to create the frame. Discreet bolted connections are used to joint the round tubular frame elements of the tower. It was important to keep a tight line to the overall structure to permit a tight fit of the screen to the exterior.

The subway entrance at the Sydney Olympic Site in Sydney, Australia, uses a curved HSS frame to create its articulated roof.

To maintain the curved lines of the tubular-framed entrance roof, discreet connections have been used. Plates are used to subdivide the diameter of the tubes to position one bolt per chamber.

The tubular frame on the exterior of the Melbourne Theatre Company building in Melbourne, Australia, designed by Ashton Raggatt McDougall Architecture, uses discreet bolted connections to connect the sections of this highly unusual geometry.

The connections use a lap joint with bolts that are recessed into holes. Care must be taken in detailing to ensure that water cannot accumulate in the connections and that bolts are corrosion-resistant.

The exterior frame on 100 Eleventh Avenue in New York City, NY, USA, designed by Ateliers Jean Nouvel, uses discreet connections to join the rectangular sections.

The connecting pieces at 100 Eleventh Avenue are carefully milled to fit with tight tolerances. The bolts are recessed into the machined holes so that the heads and nuts do not protrude.

The Joaquín Sorolla high-speed rail station in Valencia, Spain, designed by ACXT, uses a Custom AESS structure to support the train shed and arrival/departures area. The columns and beam arms of the primary (white) structure have been fabricated using brake forming as these are not standard sections – although their forms are quite clean and simple. The secondary structure that spans between the tree-columns has been painted a dark grey to allow the primary structure to be dominant.

Although the connections are obviously bolted, the arrangement of the plates is designed to be unobtrusive. The rectangular cutouts on the sides of the beams were necessary for access to the interior of the steel shapes to tighten the bolts. The holes have been simply filled by repositioning the plates and welding them in place. The welds have not been remediated.

CAST CONNECTIONS

Cast connections are used as end connectors for pin or hinge connections and as interior connections for nodes where multiple members need to be joined and the geometries and force transfer requirements may be challenging for more traditional welded and bolted methods.

Castings are used in conjunction with cable and glass structures, in complex tubular joints for buildings or in bridges. While they bring with them the added advantage of handling complex, curved geometries without the difficulties found in using multiple combinations of tubes and plates, they do require a different level of engineering and testing expertise. Economy is found in the mass production of elements. One-off castings or small runs can be very expensive. Modern steel castings are significantly higher in strength than historic cast steel as well as weldable and more ductile.

In order for castings to be cost-effective, certain conditions should be met. Travel and shipping from the foundry will be factors. Repetition of elements can amortize the cost of the mold. Nevertheless, if many elements are joining at a restricted point, or the stress through the connection is very high, then castings make sense. A rough rule of thumb is that if the connection starts to cost four times as much as the material it is made of, then steel castings start to become economical.

A cast member has a different finish. This is due to the manufacturing process and a function of the material that creates the form for the casting. If a sand casting is used, the surface texture of the finished steel will have a rough, sand-like appearance. Special finishing will be required if a seamless final appearance is sought between the casting and the adjacent tubular member as its texture will not match standard structural steel. For higher levels of AESS categories this can mean significant grinding and filling work to smoothen out the rougher finish of the casting or remove casting mill marks.

The casting process will use either an expendable or non-expendable mold. Smaller repetitive elements, such as those used in glazing and tension systems as connectors, will use a non-expendable mold and will typically be die-cast. Larger, unique, non-repetitive units will normally use an expendable mold – a mold that is broken to remove the casting. The casts are made by pouring molten metal into a mold cavity formed out of natural or synthetic sand. The sand is bonded together using clay, chemical binders or polymerized oils. The sand can be recycled to create another mold.

Castings can be formed hollow or solid. Solid castings are usually to be found in smaller connectors like the clevises that are used to form the terminuses of tension rod-type structures. Hollow castings are often used for larger members, as solid castings have difficulty in achieving uniform cooling. Non-uniform cooling can create internal stresses. Non-destructive evaluation of each large casting, including 100% ultrasonic testing, should be considered as a minimum.

Large specialty castings require specific testing by the foundry to ensure that they are properly designed and capable of resisting stresses. Cast steel exhibits isotropic properties, making it quite suitable for transferring forces through the connections in a reliable manner – i.e. being able to resist shear, moment and torsion stresses. This is achieved by manipulating the geometry as a function of variations in wall thickness, independent of the finished form of the exterior. Unlike fabrications made from tubes or plates, the interior dimensions of the void in a casting do not have to match the exterior form of the object.

The casting at Guangzhou Baiyun International Airport in Guangzhou, China, designed by URS, incorporates the pin connection at the base, the attachment of the two incoming tubular members as well as an attachment for the window system.

The support system for the tall glazed wall uses a casting to form the end termination for the vertical tubular trusses.

This highly glazed bridge between two office towers in Taikoo Place in Hong Kong, designed by Swire Properties, uses an AESS system that employs castings to maintain exceptionally well-detailed connections.

The details at Taikoo Place use a variety of approaches to pin connections, including cast clevises that have been seamlessly connected to solid rods as well as a "pinched" connection that forms the attaching point for an incoming hollow tubular member. This detail would be classed as AESS 4.

The base detail for this custom-curved support element at Stratford Station in London, England, designed by Wilkinson Eyre Architects, creates dynamic lines that provide a suitable visual termination. The use of a casting was critical for this component.

For more extensive information on cast connections and detailed information on additional projects please refer to "Chapter 10: Castings" in *Understanding Steel Design: An Architectural Design Manual.*

QRC WEST | TORONTO, ON, CANADA

Large expendable mold castings have been used to create the connections at the center of the three large 21.3m/70ft-tall delta frames that support 11 floors of concrete atop a 5 storey atrium space. The 35,000 pound castings connect eight 1m/39.375" diameter custom steel tubes. The large connectors, designed by CastConnex, were considered by the client to be the only aesthetically acceptable solution to solve the load transfer issues presented by the geometry.

Such castings push the AESS project into a Custom category directly as a result of the requirement of engineering and fabricator expertise as well as impact on the design process.

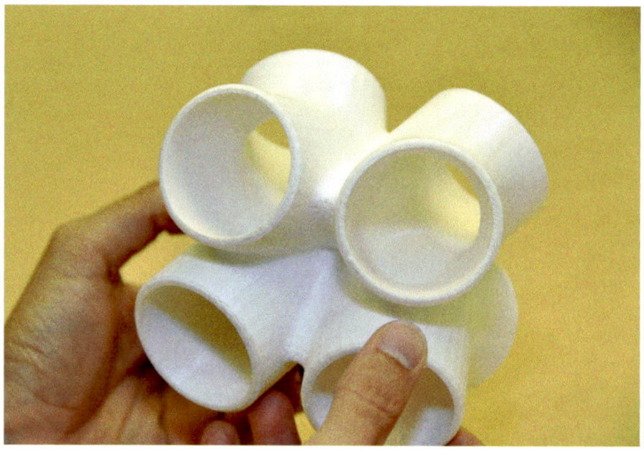

Small-scale resin models (left) were fabricated to assist in visualizing the geometry of the delta frames. This led to the aesthetic decision to use tapered cones at the base of the legs.

A small resin cast 3D printed model was used to visualize more exactly the geometry of the node (top). This type of printing technology is of great benefit to this process as a mockup is quite out of the question as these large molds are quite expensive and time consuming to fabricate.

The Tekla model produced by the fabricator, Walters Inc., shows how the castings fit into the overall structure. This type of BIM software has become critical to the design and erection processes of many AESS projects.

The casting still hot after its removal from the mold at the foundry (top left). The casting must undergo controlled cooling to avoid the creation of internal stresses.

A view towards the steel floor system (top right) that joins the top arms of the delta frames and upon which is built the reinforced concrete office tower.

After the node is cooled, the rough edges must be remediated (bottom). The desire was for a seamless appearance between the tubular legs of the delta frames and the node, so significant grinding was needed to remove the natural textured appearance of the cast node. The lugs used for the attachment of the lifting crane were cast integrally with the node.

BRISBANE INTERNATIONAL AIRPORT | BRISBANE, AUSTRALIA

The tree-like support system at the Brisbane International Airport in Brisbane, Australia, designed by Richards and Spence with Arkhefield, is set in a uniform square grid. Castings are used to simplify many of the connections.

A casting is used at the top of the strut (bottom left) to make the connection with the plate tab that has been fitted to the roof beams. In this way, some economy of repetition was made on the basis of the gridded plan for the airport layout.

The center node at the top of the gently tapered tubular column (bottom right) has been fitted with a ring to which four plates are attached to receive the tree branches/struts. The ends of the strut tubes have been fitted with castings. The weld between the casting and the tube has been left visible. This weld also serves to separate the slightly different surface textures of the casting and the tube.

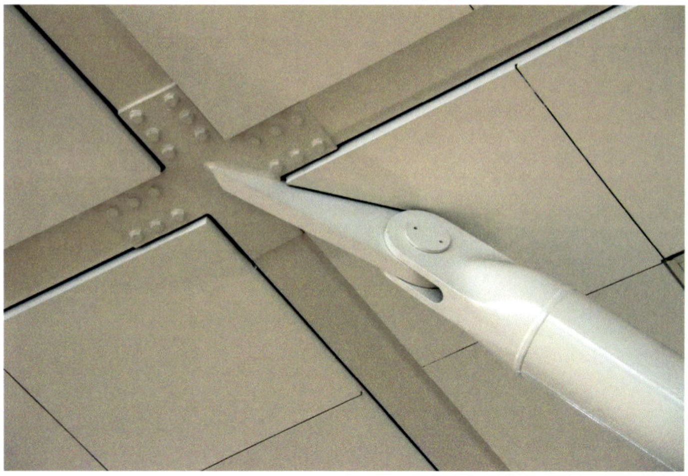

RENOVATION OF OLYMPIASTADION BERLIN | BERLIN, GERMANY

The 1936 Olympic Stadium in Berlin, Germany, was modernized in 2000 through the addition of a canopy, designed by gmp Architekten von Gerkan, Marg und Partner to provide partial cover for attendees. Although mostly cantilevered to provide the maximum of unobstructed views, the cantilever length was shortened by the addition of slender tree-like columns around the perimeter that make extensive use of custom castings.

The column supports branch into a square set of struts that are attached to the tube via custom-cast nodes.

The central node (top right) has a slightly different texture than the joining tubes. A reveal between the materials provides an enhancing detail that also serves to minimize the need for surface remediation to match the textures. In spite of the need to use an expendable (single-use) mold, there is economy in these connections through their repetition and almost identical loading requirements.

At some locations (bottom right) the node has been incorporated directly into the members and the joining butt welds have been ground and filled to make them almost invisible. This sort of treatment is more costly as it requires additional fabrication time.

REGENT PALACE HOTEL | LONDON, ENGLAND

Some large castings are based upon the aesthetics of smaller castings that are routinely used to create pin connections in tensile systems. Pin connections allow for quicker site erection as their erection process is similar to that of bolting. Typically the casting would be welded to the attached member at the shop.

The Regent Palace Hotel in London, England, designed by Dixon Jones Architects, makes use of artful custom castings to resolve the multi-member connections in this double façade.

The casting (top right) artfully directs the forces of three members of a vertical truss glazing support system into a plate connection below that receives two angled round HSS members.

The casting allows for an organic look of the connection. The transparency of the façade allows for the expression of the steel while also protecting it from weathering. The finely crafted appearance of the cast connections complements the use of a stainless steel tension support system for the glazing.

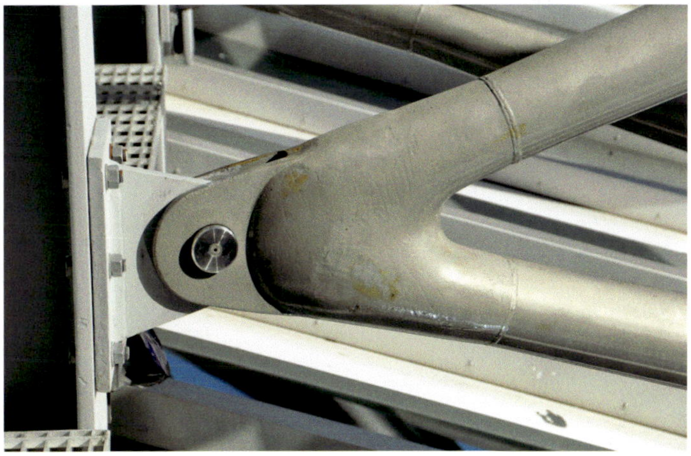

The exterior support frame on the Baltic Centre for Contemporary Art in Gateshead, England, designed by Ellis Williams Architects, uses cast connectors to make clean connections between the various tubular members.

A casting is used to connect five incoming tubes (bottom left). Slight evidence of the welded connection has been left unremediated.

This connector neatly joins two incoming tubular pieces at a pin joint. The repetition in the overall project elements helped to provide economy in the choice of castings in this project.

Custom expendable mold castings remain highly specialized and uncommon in contemporary AESS projects, yet they are used to great benefit and have proven indispensable in solving geometric and loading issues.

CUSTOM FABRICATION

CHAPTER

9

The Amgen Helix Pedestrian Bridge in Seattle, WA, USA, designed by Johnson Architecture and KPFF Consulting Engineers, required significant customization due to its widely varying geometries. Huge steel tubes twist and turn over the railroad tracks in a double helix that recalls the ladder-like structure of DNA. Although the individual members are similarly fabricated, the sizes and actual arrangement vary, adding significantly to the fabrication and erection costs.

Highly customized architecturally exposed structural steel served as the basis for the High Tech movement and continues to be a popular choice when designing contemporary AESS projects. The AESS System of Categories, Characteristics and Matrix was intended to identify highly customized fabrication of the kind that is addressed in this chapter. These projects represent the exception and not the rule. The fabrication requirements vary tremendously between projects. It will be noticed, however, that many of the detailed aspects of design and fabrication use techniques, details and approaches that draw from established practices in AESS as described in previous chapters.

The complete custom fabrication of members tends to be reserved for AESS 4 categories and "above" due to the cost premiums required for more extensive remediation associated with higher levels of welding and increased aesthetic expectations. Plate steel is often selected when standard members do not fit with the desired aesthetic or the structural capacity cannot be met with standard section sizes as referenced in *Chapter 6: Member Choices*. Where plate material is fabricated into orthogonal shapes, it provides crisp corners that may be desired as part of the aesthetic or texture of the system. Plate material can be fabricated into a wide range of tapered or irregular shapes, including curved forms.

Highly eccentric or non-repetitive geometries can require significant customization even of standard section or tube types. This approach to design is often seen in AESS 3 projects if lower levels of remediation are acceptable.

Within this chapter, custom fabrication requirements and details will be presented on a project-by-project basis.

The Beijing National Stadium ("Bird's Nest") in Beijing, China, designed by Herzog & de Meuron/Ai Weiwei/Arup, represents the extreme high end of custom fabrication. The complex geometry and overall form required the custom fabrication of all structural elements from plate material. The on-site connections were all welded, with the majority requiring significant grinding, filling and remediation to make them appear seamless.

SYDNEY OPERA HOUSE | SYDNEY, AUSTRALIA

Jørn Utzon and Ove Arup, 1973

Although steel is not likely to be the premier material that may be first associated with the Sydney Opera House, the project makes subtle use of the material as part of its refined palette. The support system for the expanses of glazing has been created from a finely detailed set of custom elements. These are comprised of steel plate that has been reinforced at its edges by the addition of round tubes. Tubular material was uncommon at the time, so this use was quite innovative. Round rods are used to provide lateral bracing between the members at their top and bottom edges.

Top left: The steel support system for the sloped glazing is carefully coordinated with the interior wood paneling. The slender braced system is anchored into the concrete structure.

Bottom left: The plate that provides the securement to the concrete structure is carefully set into the concrete to make the connection unobtrusive. This close detail shows the extreme care in workmanship and welding that has called for little to no remediation.

Right: Round steel tubes are welded to the edges of the plates. The welded "elbows" in the tubes have been ground smooth as if to infer that the tube is in one piece. The pin connections have been fabricated from rounded plates that are themselves welded to cap plates that are slightly larger than the tubes to provide an elegant detail, while at the same time simplifying the fabrication of the joint.

AMGEN HELIX BRIDGE | SEATTLE, WA, USA

Johnson Architecture and KPFF Consulting Engineers, 2004

This 128m/420ft pedestrian bridge spans over large sections of active rail track. This required the prefabrication of large sections that could be expediently erected to minimize disruption of service. The eccentric geometries in both horizontal and vertical directions required very tight tolerances to ensure fit. The deformation (camber) needed to be engineered so that gravity would pull the sections into proper alignment. The deck members were designed with heavy walled built-up plate sections. This required carefully planned weld sequences and presets to avoid heat distortion, which is often associated with this amount of extensive welding. The curved arch sections were built with high-strength steel pipe and were trial-fit section by section to ensure the final positioning would exactly match the target geometry. Fabrication and erection by Dynamic Structures.[1]

Top left: This project makes use of both standard and non-standard steel products in the creation of a dynamic-looking bridge that pushes the limits of geometry and balance. The large tubular members were rolled to curvature. The extreme length required careful splicing to at least AESS 3 standards to remove evidence of the splice in order to keep the appearance and structural nature continuous.

Bottom left: Pin connections were used to allow for the ease of erection of the smaller, truss-like members that provide the connection between the large tubes as well as support the metal mesh barriers on the sides of the bridge and the PTFE screen covering along the roof. The truss members are fabricated from welded round tubes whose end detail is made solid in appearance by the addition of a plate section that has been welded flush with the outer dimension of the tube.

Right: Tapered cones are used to mediate between the larger tubes that connect with the foundation and the smaller tubes that form the basis of the helix. Welds have been ground smooth to make the transition appear more organic.

CANOE BRIDGE | VANCOUVER, BC, CANADA

PWL Partnership Landscape Architects, 2008

The Canoe Bridge was part of the improvements to the False Creek area of Vancouver in preparation for the 2010 Olympic Winter Games. Its shape was intended to evoke that of a canoe, using a predominantly welded tube system that incorporated custom curved, standard and tapered members. The decking is grated and the bridge open to allow maximum light to the marine habitat below. The 40m/131ft bridge has one central V-support to minimize environmental impact and allow the individual spans to keep a low profile. Fabrication by Megatech and erection by Wilco.[2]

Top left: The bridge was erected as two prefabricated elements. The V-support system was first installed. The bridge element arrived on the site in one piece and was lifted by crane onto the support system.

Bottom left: The outer custom-curved tubes of the bridge are tied back to the walkway and railing system. The tapered members that tie back to the railing are custom-tapered sections with bolted attachments at the distant ends to provide an enhanced detail that is intended to contrast with the all-welded nature of the balance of the bridge structure.

Top right: The project uses the shape of the steel material to differentiate its systems. Tubular steel (white) is used for the primary structure of the bridge and plate material (grey) for the railing system. The railings are tied to the primary structure, providing additional stability to the 3D truss sections that frame the pedestrian path.

Bottom right: The supports consist of custom-tapered tubular elements. The view from below shows the use of a triangular truss system of all-welded tubular steel that constitutes the support deck for the walkway.

VIADUCT EVENTS CENTER | AUCKLAND, NEW ZEALAND

Moller Architects, 2011

The roof form of the Events Center was derived from the need to provide large column-free space. Cutting and shaping the steel was done manually by D&H Steel through the use of scallop cutting templates to cope with the many shapes. The roof sections were shop-assembled on a 42m x 7m/138ft x 23ft jig. In response to the harsh marine environment the steel was given a protective zinc coating under the final white finish. Each truss is 40m/131.2ft long, 7.5m/24.6ft wide and weighs 55 tonnes.

Top left: A view to the underside of the undulating roof form shows the expression of the tubular steel diagrid structure. The columns are tapered, narrowing at both the top and bottom points of attachment to increase their slender appearance. Curved steel plate provides additional stability and forms the finished surface to the underside of the canopies on the exterior.

Top right: On the interior, the ceiling is finished with an acoustic wood-based material, which succeeds in highlighting the presence of the steel in the space.

Bottom left: The barrel-vaulted roof segments were prefabricated to allow for lifting onto the structural frame as a completed unit. The curved diagonal-grid tubular frames were connected to the curved plate-steel roof sections, including the pre-attachment of the large curved plate-steel gutter elements.

Bottom right: The extensive use of curved elements requires a customized view to the fabrication and detailing of the building. All-welded connections and in-shop fabrication assisted in maintaining tight tolerances and geometric regularity. The welds have been neatly done and left unremediated.

THE SHARD OBSERVATION LEVEL | LONDON, ENGLAND

Renzo Piano Building Workshop and WSP, 2013

The observation level of The Shard makes subtle use of custom plate steel to create its relatively technical-looking exposed steel structure. Hot-rolled members would not have provided the crisp edges, so the sections are fabricated from welded plate; the corner elements are square and the diagonals mimic a wide-flange (Universal) shape. In many of the end connections of the bracing members the flanges have been rolled to create a subtle custom look. Fabrication and erection by Severfield Reeve Structures.

Top left: The subtle use of custom plate steel has been showcased here. The desire was for sharp edges of the exposed structure as well as hollow square columns at the corners and diagonal braces that appear as wide-flange (Universal) members. The white finish on the steel contrasts with the exposed grey services above.

Bottom left: The impression of the use of wide-flange members creates a more industrial aesthetic to the space. The steel used to support the glazing system uses different custom shapes to create the clear sense that the envelope is a distinctly separate system.

Top right: This detail uses plate material in order to create the gently curved connection, sharp edges and inset on the square column to allow for the "nesting" of the flange of the incoming diagonal member.

Bottom right: The AESS on the interior of the viewing area has been modified to remove the technical feel of the space through the creation of hollow rectangular sections whose connection detailing has been suppressed.

THE NEW YORK TIMES BUILDING | NEW YORK CITY, NY, USA

Renzo Piano Building Workshop, FXFOWLE Architects and Thornton Thomasetti, 2007

The design for the New York Times Building had as its aesthetic intention to reflect the permeability and transparency of the relationship between the newspaper and the city. To this end the exterior AESS structure supports a layer of the double façade that consists of a lighter steel frame, which in turn supports horizontal ceramic rods that act as a sunshade and a heat sink. The bracing of the steel frame takes a key role in the building's aesthetic expression at the recessed corners that mark the discontinuity of the screen elements.

Top left: The double façade uses an expressed AESS bracing system on the exterior. AESS is additionally used to support the exterior screen skin of the double façade.

Bottom left: AESS is used to create a textured, multi-layered façade system reminiscent of Piano's earlier High Tech design, while being more subtly contemporized to the highly urban New York site. The grey color of the steel is well suited to be durable.

Top right: Although at first glance the members look like standard sections, their sharp corners and attachment systems reveal that they were custom-fabricated from plate steel. Although not all welds have been ground and blended, the nature of the structural design as well as its integration of a tension bracing system would place this in an AESS 3 category.

Bottom right: Custom castings have been used to create the support system for the double façade. These incorporate a bolted system of connections to make on-site assembly more efficient.

NATIONAL ARCHIVES OF CANADA | GATINEAU, QC, CANADA

Blouin IKOY & Associés, 1996

The project had an extremely high mandate for durability and lowered maintenance to answer to the task of protecting the important documents that it houses. The structure has been designed to last 500 years. It used stainless steel to fabricate the custom 24.4m/80ft-tall columns that are symmetrically placed with the expressed purpose of resembling a temple. Corrosion-resistant Type 316 L stainless steel (UNS 31603, EN 1.4404, SUS 316L) was used for all structural components outside the building envelope. The steel was given a glass bead blasted finish.[3]

Top left: The building is somewhat unusual for an extremely cold climate zone in that the structure is placed on the exterior of the enclosure system. The use of stainless steel is meant to contribute to long durability.

Top right: The project makes use of a combination of custom-fabricated and standard structural shapes to create its unique aesthetic. This is characteristic of the earlier work of IKOY, which often used enhanced detailing to add value to fabrications that used standard structural shapes.

Bottom left: This detail of a tension connector shows the extreme care in the fabrication detailing for the structure. The selection of stainless steel places higher requirements on the fabrication process.

Bottom right: The detailing of the base of the tower elements for the columns exaggerates the connections by using oversized elements that have been arranged in a manner that clearly separates the structural functions of the components.

CANNON STREET STATION AND OFFICE BUILDING | LONDON, ENGLAND

Foggo Associates, 2012

The structure of the Cannon Street Underground Station redevelopment needed to respond to some difficult site conditions. Situated above the rail lines and on a sensitive archeological site, the foundation points were very limited. The building was designed for a 21m/69ft-deep strip of the block and cantilevered from the restricted support position. The north and south façades work as large 67.5m/221.5ft-long truss frames, comprised of round HSS with diagonal braces. These are separated from the large X frame that has been fabricated mostly from steel plate with thicknesses of up to 100mm/4in. Steel fabrication and erection by Watson Steel Structures Ltd.[4]

Left: Several smart moves have been incorporated to simplify the fabrication. In lieu of aligning the plate steel to create true corners, the face steel on the diagonals has been recessed to permit fillet welding and obviate the need for grinding. Similarly, the end caps on the square tubes that brace between the members protrude slightly, again allowing for fillet welding and a small size of this detail. The square tubes are connected via a discreet and tight connection that also creates an attractive shadow line. Allowing adjacent surfaces to be offset eliminates the need for butt welding, grinding and filling of surfaces.

Top right: The building structure also clearly expresses the support of its cantilevered components through the differentiation of the massive cross-bracing detail on the side elevation, featuring the more finely grained, diagonally braced tubular system of the primary façade.

Bottom right: Although the overall appearance of the exterior structure is highly customized, the primary frame uses standard round HSS members that are joined via in-shop welding for the tube intersections and also via discreet connections for site erection. The cast connections that terminate the diagonal braces are easily formed to rounded plates via pin connections. The rounded plates have been simply welded to the tubular frame. The repetition of the details throughout the façade affords some economy.

PUENTE DE LUZ | TORONTO, ON, CANADA

Francisco Gazitua, 2012

Pedestrian bridges are serving to reconnect areas of cities that have been divided by rail corridors and highways. More recently they are being elevated to serve as public art. The design and fabrication of this bridge required significant collaboration between the artist and the steel fabricator Walters Inc. The use of a pedestrian bridge as a piece of public art often allows for a higher level of funding. Of course, pedestrian safety and durability will remain the main factors of the design. Here the bridge elements were fabricated off-site and lifted with two significant lifts, facilitated by the positioning of the central concrete pier. This resulted in minimal disturbance to the commuter rail tracks.

Top left: This downtown pedestrian bridge reconnects two pedestrian precincts that are separated by a number of active train tracks.

Bottom left: A view down the bridge shows the contrast in the selection of galvanized mesh materials for the fall protection barrier with the painted steel for the structure of the bridge. The placement of the mesh serves to protect the edges of the steel from damage and wear due to traffic. Although the mesh is quite dense to prevent climbing, it still permits reasonable views through to the city beyond.

Top right: The detailing of the steel structure is intentionally playful. The curved hollow box sections are created from plate that has undergone brake forming. The faces of the box sections are offset to permit more simple welding and finishing, in correspondence with AESS 3 standards. Galvanized bolts provide a durable finish and express the on-site connections. The project makes a feature of the contrast between its fabrication strategies.

Bottom right: Simple strategies have been employed in the design of the connections. Sizing the end plate for the tubes larger than the diameter of the tubes allows for fillet welding and obviates the need to grind the connections. This is true also for the plate that serves to connect the two angled tube members. The roundness of a precision-cut plate is more accurate than the circular tubes that are formed by rolling processes. This means that making a precise, invisible cap plate on the end of a tube is difficult to achieve. The larger the diameter, the more likely there will be a geometric mismatch. This would represent excellent AESS 3 detailing, where good choices have minimized the need for grinding.

ARGANZUELA BRIDGE | MADRID, SPAIN

Dominique Perrault Architecture with MC2 Estudio de Ingeniería, S.L., 2010

This footbridge connects pedestrians to a newly developed park alongside the Manzanares River. The diameter of the circular cross-section of the bridge varies from 12m/39.3ft at the center to 5m/16.4ft at the entry points. The bridge is designed as two independent sections that open into the central park area. The span lengths are 150m/492ft (with a main span of 115m/377ft over the river) and 128m/420ft (with a span of 96m/315ft over the park). The custom-fabricated curved steel box sections have been designed as two intersecting spirals. The construction is technically described as a spatial truss with circular cross-section, whose diagonals are curved rectangular tubes inscribed in a tronco-conical surface.[5] A ribbon of steel mesh winds down the length to protect patrons from the sun. Alternating segments use more simple plate steel sections to provide partial shade and increased ventilation. The combination of materials creates a remarkable quality of light both inside and outside the bridge.

Top left: The main span of the bridge could not be created in one lift due to the size and weight of the structure as well as limited accessibility by cranes. The bridge elements arrived in partial sections and the diagonal side ribs were welded on site. Limitations of the site conditions fed heavily into the need for on-site welding.

Bottom left: The large variety of connection types form part of the purposeful aesthetic language of the bridge. As much on-site welding was necessary, the decision was taken to leave the welds exposed and unremediated.

Right: The interior space of the bridge is defined by the changing qualities of light as transmitted through the varieties of screens and meshes. The welded connections at the corners of the box sections and at the connection points between the side ribs and the top and bottom of the tube sections are left apparent.

ORIENTE STATION | LISBON, PORTUGAL

Santiago Calatrava, 1998

Constructed for Expo '98, the Oriente Station project includes bus and rail terminals as well as access across a major roadway to a shopping center that now forms the entrance to the former Expo site. The bus terminal and large steel and glass roof atop the rail platforms make intense use of components that have been custom-fabricated from plate material. The project, though highly complex in appearance and predating contemporary software, makes use of repetition of elements to gain some economy in prefabrication.

Top left: The main façade is highlighted by a central curved canopy that cantilevers over the passenger drop-off area. Two pedestrian bridges create upper level links on either side of the canopy. The curvature of the foreground elements is contrasted by the angles of the rail canopy.

Bottom left: Long cantilevered canopies provide cover for the bus loading area. The primary elements are comprised of custom-fabricated plate steel. These are structurally reinforced in the fashion of a wide-flange (Universal) section through the addition of flange elements. The flanges of the structure were curved using brake forming.

Top right: A view from the upper level of the bus terminal back towards the angled roof of the train platform. Custom fabrication is needed to respond to the continuity of the flowing form of the canopy supports for the bus terminal as they rise to become the support system for the roof of the upper level of the station.

Bottom right: A detail of the rail station roof showing the adaptation of the wide-flange section into a series of branches whose curved, continuous intersections make use of brake forming to create the curvature in the flanges.

PEACE BRIDGE | CALGARY, AB, CANADA
Santiago Calatrava, 2012

The Peace Bridge has a span width of 126m/413ft and an overall width of 8m/26ft. It is constructed as a helical steel structure. Sitting low to the river, it was designed to be able to resist a 100 year flood, an event that happened only a year after its opening. The design of the bridge deviates significantly from other Calatrava designs in the absence of a mast and cable system. The color of the bridge was chosen to reflect the color both of the Canadian and Calgary flags. It is a highly successful combination of the function of the pedestrian and cyclist connection and public art for the city.

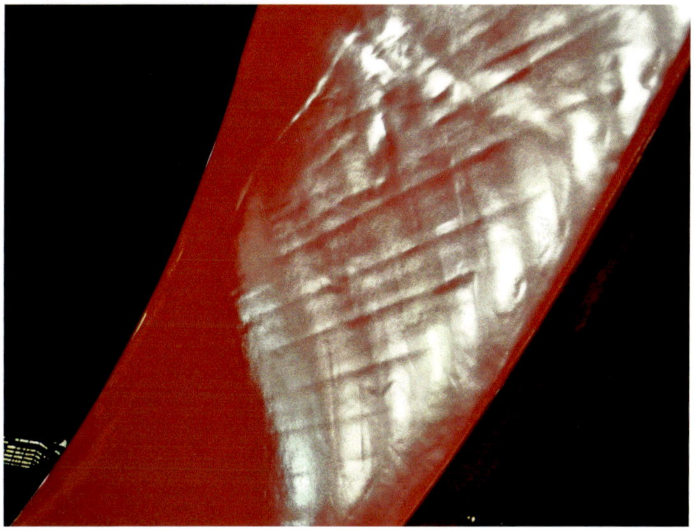

Top left: The Peace Bridge shows a very slight curvature as it provides near-level passage across the Bow River for pedestrians and cyclists.

Bottom left: The interior of the passageway is partially protected by a translucent covering. The ribs have been fabricated from double curvature bent plate, with the welded connections fully remediated by grinding and filling. Their complex curvature succeeds in creating a plastic feel to the steel that places its fabrication even above AESS 4 and into a highly custom range due to many additional requirements to achieve this aesthetic.

Top right: The bridge has a different appearance at night, as the lighting now comes from within the structure. The differences between the day and night conditions of light should always form part of the design strategy when considering color, finish and the selection of lighting. The coloration of the light source will impact the reading of the color of the paint, understanding that over time and with the introduction of new, more energy-efficient light sources, this may be a changing parameter.

Bottom right: Although the fabrication method is well hidden during the daylight hours, night lighting from within the bridge makes evident the brake forming method. The complex curvature of the ribs has required that brake forming occurs as an overlapping pattern in order to create these complex curves.

GATESHEAD MILLENNIUM BRIDGE | NEWCASTLE, ENGLAND

Wilkinson Eyre Architects with Gifford, 2001

This innovative bridge was designed to fully rotate in order to permit the passage of boats beneath, while in its normal position provide close-to-level passage across the River Tyne for pedestrians and cyclists. It is entirely fabricated from custom plate material that makes extensive use of curving processes such as brake forming. The structure was fabricated off-site and delivered by barge, being lifted into place in one piece by one of the world's largest floating cranes, the Asian Hercules II.

Top left: The bridge in its closed position, showing the relationship to the river and the relative flatness of the crossing path.

Bottom left: The crossing path is attached to the main arch via a series of simple tension elements. The three elements – arch, tension rods and curved path – create distinct statements arising from the nature of their fabrication and purposeful use of steel.

Top right: The welded connection between the large arch and the rotating base shows the use of curvature in the plate as well as the simple use of unremediated welds. This provides a clear structural reading between the two elements that would become less clear with a blended fabrication treatment.

Bottom right: The curvature of the pedestrian walkway and the tension support system provide a geometry that permits the thin profile of the walkway that is constructed for the most part from reinforced custom-curved steel plate.

PEDRO AND INÊS BRIDGE | COIMBRA, PORTUGAL

Cecil Balmond, 2006

The Pedro and Inês Bridge that spans over the Rio Modego is 274.5m/900.6ft long and 4m/13.1ft wide. The primary structure is fabricated from large sections of plate steel. The structure is challenged by an offset of the two sides of the bridge at midspan. It was able to be erected sequentially from either side of the river, using the cantilever design from the support points to provide the stability until the central plaza could be completed and provide ultimate stability. Steel plate sections are welded within the box forms to provide a hidden structure. Care needed to be taken to prevent show-through from the heat of the welding in order to keep the nature of the interior structure from view.

Top left: A view to the underside of one of the bridge support points. The plate steel form was designed to provide the bridge with a very organic yet geometric appearance that belies the traditional aesthetic of a steel bridge. The corner conditions are well finished to mask traces of fabrication.

Bottom left: The walking surface of wood decking forms a contrast to the steel and colored glass railing system.

Top right: The faceted railings of the bridge have also been fabricated from sections of plate and flat bars. The glazing is held in place with a system of thin steel disks welded to the railing members, in the creation of an obvious aesthetic of fastening.

Bottom right: The underside of the bridge reveals unremediated welds between large segments prefabricated from steel plate. The light cast on the surface gives slight glimpses of the weld show-through of the steel ribs inside of the structure.

ABC MUSEUM | MADRID, SPAIN

Aranguren + Gallegos Arquitectos, 2010

The trusses that support and enclose the café that forms the entrance to the museum courtyard make innovative use of plate steel to create highly irregular patterns. The light load capacity of the individual plates is accommodated by the density of the diagrid pattern.

Top: The structural façade uses AESS welded plate steel to create an irregular diagonal grid truss that spans between the opposing buildings thereby framing the courtyard.

Bottom Left: Although the detailing of the welded connections appears simple, there were challenges with the acute nature of the angles and the maintenance of the visual lines, given that many of the plate sections are discontinuous. With some weld remediation the structure would be classed at AESS 3.

Bottom right: The welded plate-steel diagonal grid creates the artistic expression of the glazed link that comprises the museum café. Although the individual members are thinner than would be expected for a vertical structure, plate steel as a system is able to provide a stable truss. Again, certain of the diagrid elements are chosen to be continuous; the intersecting pieces are sized with low tolerances to ensure good contact for fillet welding and alignment to provide an appearance of continuity.

ST. JOHN AMBULANCE HEADQUARTERS | EDMONTON, AB, CANADA

Manasc Isaac Architects, 2004

Plate steel was used in a highly unusual way to create an integrated column-and-“beam” system for this atrium space. The V-shaped columns divide at the top to create single-thickness plate members that taper to minimal thickness where they are connected to the wide-flange columns that support the glazing system. Butt welds used to splice the plate segments have been fully remediated and made to disappear.

Top left: In working towards a LEED® Silver sustainability qualification, the building sought to achieve high levels of daylighting. The use of steel in the multi-storey glazed atrium was part of that objective.

Bottom left: The selection of custom plate members that mimic a more regular wide-flange (Universal) section to support the glazing wall acts in contrast to the highly unique plate elements that support the interior wall. HSS have been used to brace between the window members.

Top right: A view directly up the column shows the simple use of plate steel on edge to create the support system for the roof of the multi-storey atrium. The bent plate sections are connected intermittently up the center. A continuous connection was structurally undesirable. The angled cuts with plates connect the two sections, while simultaneously adding no stresses to the connections of the very tall members.

Bottom right: Custom plate columns allow for the tapering of the section towards the lower portion of the glazing to provide a more elegant base condition.

BOSTON SOCIETY OF ARCHITECTS STAIR | BOSTON, MA, USA

Höweler + Yoon Architecture, 2011

This stair, which is the central feature of a renovated space to house the Boston Society of Architects, bridges between requirements of AESS structures and miscellaneous metals in terms of logistics, size, weight and fabrication issues. The design of the stair intended to have it appear to be constructed from a single, flowing element of plate steel. As the stair needed to be subdivided into separate pieces for transportation and lifting, discreet connections were designed to give the appearance of continuity. The fabrication relied on brake forming to create the bends in the treads and railing elements.

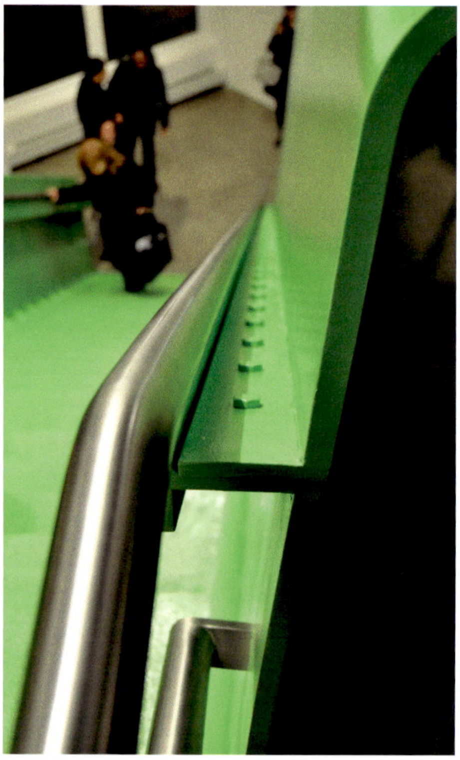

Top left: The stair unit was fabricated off-site and shipped as one single element. Logistics included the need to fit the stair through a limited opening in the front wall of the existing building as well as to manipulate a crane to perform the lift. The stair is hung from the structure by large sections of plate. The joints in the individual plate sections are barely visible as vertical lines, thanks to the tight tolerances observed in the fabrication.

Top right: The connection between the large plate section used to hang the stair from the structure, and the stair itself (which was erected as one unit) is created with a simple bolted detail that unobtrusively sits behind the stainless steel handrail. This allowed for ease of erection and avoided difficult on-site welding.

Bottom: The treads have been created as brake-formed U-shaped members. They have been connected to large plate "stringers" via butt welding. The welds have been ground smooth and fully remediated to leave the connection intentionally mysterious.

CAIXA FORUM | MADRID, SPAIN

Herzog & de Meuron, 2007

The Caixa Forum required innovative use of steel in order to create a specialized structural system for the first floor of the renovated historic building, so as to provide the appearance that the existing masonry structure was floating above the public plaza. Deep steel beams span between vertical support points deeper into the plan of the building so that the support points are effectively set back into the shadows and hidden from view. The floor is suspended from AESS hangers, the underside of which consists of an exposed, undulating, welded steel plate system.

left: Caixa Forum makes use of weathering steel for the upper levels of its façade, while plate steel creates the underbelly of the public space that is formed as the renovated historic masonry structure is suspended from the new interior support system.

Top Right: A view of the underside of the large cantilever reveals the unremediated butt weld joints that have been used to connect the steel plate. The uneven geometry uses the ability of the plate to subtly deviate from being absolutely planar.

Bottom right: Support beams span between discreet load-bearing vertical elements in the museum to enhance the feeling of openness that also accentuates the infrequently spaced vertical hangers used to suspend the cantilevered floor system. Vertical round HSS members are used to hang the floor from the beams above. To assist in stabilizing the hangers, triangular plate steel elements have been welded between the top of the hangers and the steel beams. This makes for a stiffer, more moment-resisting connection and also adds visual detail. Note also the web stiffener that has been welded to the beam at this point of connection to assist in load transfer.

The use of the primary AESS 1, 2, 3 and 4 categories leave the designation of Custom work to projects that truly warrant such detailing, thereby assisting in lowering the costs of AESS projects and increasing the viability of using AESS. The specific treatment of detailing with the type of workmanship associated with fabrication methods such as grinding and contour blending for categories AESS 4 and above can help prevent unnecessary over-specification of fabrication methods that are not warranted due to the use of the building, the distance to view and the type of finish required.

The application of Architecturally Exposed Structural Steel in a project can greatly benefit by the careful use of the system of categories and their associated characteristics. This system is able to create a much cleaner decision-making matrix and assist in creating better communication between the architect, engineer and fabricator regarding the expectations of the design of the exposed steel in the project.

The Phoenix International Media Center in Beijing, China, designed by BIAD Ufo, showcases the use of AESS in conjunction with a high level of parametric design. A deep understanding of the structural capabilities of steel as well as effective detailing and fabrication methods were essential for the success of this project.

SELECTED BIBLIOGRAPHIC REFERENCES

Boake, Terri Meyer. *Understanding Steel Design: An Architectural Design Manual*. Birkhäuser, 2011.

—. *Diagrid Structures: Systems, Connections, Details*. Birkhäuser, 2014.

HIGH TECH ARCHITECTURE

"Architecturally Exposed Structural Steel: A Design Guide." *Modern Steel Construction*, May 2003.

Banham, Reyner. "Art and Necessity: Inmos and the Persistence of Functionalism." *Architectural Review*, December 1982.

Buchanan, Peter. "Essay: High-Tech." *Architectural Review*, June 6, 1983. http://www.architectural-review.com/home/ar-archive/ar-july-1983-essay-high-tech-by-peter-buchanan/8604479.article

Boake, Terri Meyer. "Architecturally Exposed Structural Steel: How Is It Defined?" *Advantage Steel*, Spring 2005.

—. *CISC Guide for Specifying Architecturally Exposed Structural Steel*. Canadian Institute of Steel Construction. May 2011. http://www.cisc-icca.ca/solutions-centre/aess

—. "Form, Fit and Finish: Design and Specification of Architecturally Exposed Structural Steel.." *Construction Business*, July/August 2012.

—. "The Art of the Bridge." *Advantage Steel*, Winter 2014.

Boulanger, Sylvie, and Terri Meyer Boake. "The Canadian Matrix: A Category Approach for Specifying AESS." *Advantage Steel*, Summer 2008.

—. "The 3 Cs of AESS." *Modern Steel Construction*, December 2011.

Boulanger, Sylvie, Terri Meyer Boake, and Walter Koppelaar. "A Categorical Approach: Specifying AESS." *Modern Steel Construction*, April 2008.

Kloiber, Larry. "Designing HSS Connections." *Modern Steel Construction*, November 2001.

BENDING STEEL

Alwood, Todd. "What Engineers Should Know About Bending Steel." *Modern Steel Construction*, May 2006.

"Bending Steel: Questions Answered." *Modern Steel Construction*, June 2002.

CASTING STEEL

Baddoo, Nancy R. "Castings in Construction SCI Publication 172." Online Information Center for Stainless Steel Construction. February 2008. http://stainless-construction.com/ss_info_cntr/pdfs/Castings%20in%20Construction.pdf

Boake, Terri Meyer. "Branching Out." *Modern Steel Construction*, July 2008.

de Oliveira, Carlos, and Tabitha Stine. "Convenient Connections." *Modern Steel Construction*, July 2008.

de Oliveira, J.C., S. Willibald, J.A. Packer, C. Christopoulos and T. Verhey. "Cast steel nodes in tubular construction - Canadian experience." *Proceedings of the Eleventh International Symposium on Tubular Steel - ISTS 11, Quebec City*, 2009.

"Non-Expendable Casting Techniques." *Castings and Forgings News*. http://www.castings-forgings-news.com/castings/ShowPage.aspx?pageID=1572

Schober, Hans. "Steel Castings in Architecture and Engineering." *Modern Steel Construction*, April 2003.

Stuart, D. Matthew. "Wrought and Cast Iron Structures." *Architectureweek*. June 15, 2011. http://www.architectureweek.com/2011/0615/building_2-1.html

CONCRETE FILLED STEEL TUBES

Schneider, Stephen P., Donald R. Kramer and Douglas L. Sarkkinen. "The Design and Construction of Concrete-Filled Steel Tube Column Frames." *13th World Conference on Earthquake Engineering*, 2004.

ELLIPTICAL TUBES

Packer, Jeff. "Elliptical Hollow Sections Part One: Properties and Applications." *Advantage Steel*, Fall 2009.

—. "Elliptical Hollow Sections Part Two: EHS Member Design." *Advantage Steel*, Winter 2009.

—. "Elliptical Hollow Sections Part Three: EHS Connection Design." *Advantage Steel*, Spring 2010.

—. "Going Elliptical." *Modern Steel Construction*, March 2008.

COATINGS, FINISHES AND FIRE PROTECTIVE DESIGN

Alfawakhiri, Farid, Christopher Hewitt and Robert Solomon. "Fire Protection: Common Questions Answered." *Modern Steel Construction*, December 2002.

"Bird's Nest - China." http://www.galvanisationzinga.com. http://www.galvanisationzinga.com/pdf/realisations/autres/Bird%27s_Nest_en.pdf

Burgan, Bassam. "Guidance on the Use of Stainless Steel in Construction." Edited by The Steel Construction Institute, UK. *Symposium Structural Applications of Stainless Steel in Buildings and Architecture*. Brussels: Euro Inox, 2000.

Ferguson, Anthony, and Barbara Lane. "Arup Fire." *Swiss Re – structural fire engineering for the newest tall building in London*. http://www.mace.manchester.ac.uk/project/research/structures/strucfire/CaseStudy/steelComposite/swiss.htm

Fire Protection of Plain Concrete Filled Hollow Steel Structural Sections. Bulletin No. 21, Canadian Steel Construction Council, 1994.

Gewain, R.G., N.R. Iwankiw, F. Alfawakhiri and G.S. Frater. *Fire Facts for Steel Buildings*. Canadian Institute of Steel Construction, 2006.

"Hot Dip Galvanizing for Sustainable Design." American Galvanizers Association. 2009. http://www.galvanizeit.org/images/uploads/publicationPDFs/Galvanizing_for_Sustainable_Design.pdf

Kodur, V.R. "Solutions for enhancing the fire endurance of HSS columns filled with high strength concrete - NRCC- 47362." National Research Council Canada. January 2006. http://www.nrc-cnrc.gc.ca/obj/irc/doc/pubs/nrcc47362.pdf

Parker, Arthur J., Jesse J. Beitel and Nestor R. Iwankiw. "Fire Protection Materials for Architecturally Exposed Structural Steel (AESS)." *Structure Magazine*, April 2005.

Pollack, Beth. "Fire Protective Coatings: An Overview." *Modern Steel Construction*, March 2005.

"Zinc Coatings." American Galvanizers Association, 2006. *http://www.galvanizeit.org/images/uploads/publicationPDFs/Zinc_Coatings.pdf*

NOTES

CHAPTER 1

1 Buchanan, Peter. "Essay: High-Tech." *Architectural Review*, June 6, 1983. http://www.architectural-review.com/home/ar-archive/ar-july-1983-essay-high-techby-peter-buchanan/8604479.article

CHAPTER 2

1 The Society for Protective Coatings. http://www.sspc.org/about-us/about-us-what-is-sspc/

CHAPTER 3

1 The Society for Protective Coatings. http://www.sspc.org/about-us/about-us-what-is-sspc/

CHAPTER 5

1 Stratford Town Centre Link. http://www.designingbuildings.co.uk/wiki/Stratford_Town_Centre_Link

2 "A Walk in King's Park." *Architecture AU*. http://architectureau.com/articles/a-walk-in-kings-park/

3 Clayton International Industrial and Commercial Fireproofing: Experience Music Project. http://www.claytonnational.com/index.php/music-experience-project.html

CHAPTER 9

1 Empire Industries Limited: Amgen Helix Pedestrian Bridge. http://www.empind.com/portfolio/amgen-helix-pedestrian-bridge-seattle-wa/past-projects

2 The Challenge Series: Millennium Water: The South False Creek Olympic Village, Vancouver, Canada. http://www.thechallengeseries.ca/chapter-03/parks-waterfront/

3 http://www.ikoy.com and International Molybdenum Association, http://www.imoa.info/molybdenum-uses/molybdenum-grade-stainless-steels/architecture/sustainability-illustrations. php

4 Steel Construction: Cannon Place. http://www.steelconstruction.org/resources/design-awards/2011/award/cannon-place-cannon-street-station-london.html

5 MC2 Estudio de Ingeniería: Arguanzela Footbridge. http://www.mc2.es/ficha_proyecto.php?id=67&idcategoria=8&i=en&nav=02

ILLUSTRATION CREDITS

All photographs are by the author, unless noted otherwise.

Cover: The World Financial Center entry pavilion isometric, Walters Inc.

22 QRC West detail, Walters Inc.
44 Intermittent weld caulking, Sylvie Boulanger
47 Ottawa International Airport, isometric drawing, Walters Inc.
48 Pearson International Airport, mock-up, Walters Inc.
48 Calgary International Airport, mock-up, Sylvie Boulanger
65 Ontario College of Art and Design, transport of legs, Walters Inc.
67 Union Station, isometric, Walters Inc.
68 Ontario College of Art and Design, temporary shoring, PCL Constructors
71 Capital Gate, Miroslav Munka
72 Canton Tower Construction (3), Arup
77 Guelph Science Building, cast node, Walters Inc.
150 QRC West, delta frame models, CastConnex
150 QRC West, isometric, Walters Inc.
151 QRC West, hot node, CastConnex
151 QRC West, grinding the node, CastConnex
161 Viaduct Events Centre, lifting the roof, D&H Steel
184 Photo of author, Ian Godfrey

SUBJECT INDEX

INDEX OF BUILDINGS

INDEX OF PERSONS AND FIRMS

ABOUT THE AUTHOR

Terri Meyer Boake BES, B. Arch, M. Arch, LEED AP

Terri Meyer Boake is a Full Professor at the School of Architecture at the University of Waterloo, Cambridge, Ontario, Canada. She is Past President of the Society of Building Science Educators – an international group of academics who teach in the area of sustainable design. She is also Past President of the Building Technology Educators' Society, a group dedicated to vitalizing teaching in the areas of structures and construction.

She has been teaching in building construction and structures since the early 1980s and launched an effort to expand the environmental core curriculum at the School of Architecture at Waterloo during the mid-1990s. She works with the Canadian Institute of Steel Construction (CISC) and the Steel Structures Education Foundation of Canada to develop interactive multimedia case studies to promote the teaching of steel construction in schools of architecture across Canada. She worked with the CISC Architecturally Exposed Structural Steel task force and has published the *CISC Guide for Specifying Architecturally Exposed Structural Steel* in 2011. She worked to adapt the CISC AESS Guide for the Australasian market and has toured Australia and New Zealand to promote the system. She is presently working with the American Institute of Steel Construction to expand the use of the CISC AESS methodology to the USA. She lectures extensively on Architecturally Exposed Structural Steel. Much of this research was used to inform her recent books *Understanding Steel Design: An Architectural Design Manual* and *Diagrid Structures: Systems, Connections, Details*. Her ongoing intense research and investigations into AESS have been essential to the development of this current book. This will be further developed in her next book publication *Complex Steel Structures*, which will provide a continued investigation of issues of non-orthogonal geometries.

Terri Meyer Boake holds a professional degree in Architecture from the University of Waterloo and a post-professional degree in Architecture from the University of Toronto. She is a LEED Accredited Professional in sustainable building. Actively engaged in researching carbon-neutral building design, with an emphasis on architectural and passive solutions over a reliance on system and energy sources, a current research project involves the creation of resource materials to assist educators and practitioners in the better understanding of how to create carbon-neutral buildings.

She is an avid photographer and has spent significant time traveling to document and bring noteworthy buildings to the attention of the community. Noting a deficiency and lack of critical construction process information available in current publications, her goal has been to document and present buildings with a "construction eye".

She is an active member of the Council on Tall Buildings and Urban Habitat (CTBUH) and of the Board for the Skyscraper Center (www.skyscrapercenter.com), a database of information and images of tall buildings from around the world. One of the main focuses of this work is the photographic documentation of tall buildings and work with the Height Committee to refine the current definitions for the structural systems of tall buildings.